A History of the Northeast Algal Society (NEAS)

by

H.William Johansen & Peter M. Bradley

Bill Johansen
Department of Biology
Clark University
Worcester, MA 01610
Hjohansen@clarku.edu

Peter Bradley
Department of Biology
Worcester State University
Worcester MA 01602
Pbradley@worcester.edu

NEAS 50th Anniversary logo, created by Bridgette Clarkston and Mike Wegner, using *Licmorpha* drawing from Ernst Haeckel's *Kunstformen der Natur* (Artforms of nature) (1904).

ISBN-13: 978-1460973202

ISBN-10: 1460973208

NEAS Contribution Number 4

TABLE OF CONTENTS

Contents

LIST OF FIGURES

Figures continued

INTRODUCTION

What better time to prepare a history of an organization than on its half century birthday. With the fiftieth anniversary symposium of the Northeast Algal Society (NEAS) scheduled for the weekend of 15-17 April 2011, it behooves us to gather together minutes, programs, memos and other relevant documents from which a history of the organization can be assembled. Hopefully, this document will show how NEAS evolved since its origin in the 1960s.

For two major reasons the fiftieth anniversary celebration of NEAS should prove memorable. First, this is NEAS's fiftieth birthday and, second, the symposium will be held at the Marine Biological Laboratory in Woods Hole, Massachusetts, a venue where many of the NEAS symposia in the 1970s, 1980s, and 1990s were held. However, from 1998 until 2010, the symposia have been convened at conference centers elsewhere, so this will be like a homecoming, at least for senior members of the society. As printed in the first circular for the 2011 symposium, NEAS is returning to "…the roots of our society."

Now, the fiftieth birthday of NEAS is an excellent time to update an older document on the history of NEAS that we assembled several years ago (Johansen & Bradley 2003). By collecting and analyzing the many documents prepared during these years relating to NEAS, we probably have the pulse of the organization. Also helpful in writing this report is the fact that one of us is emeritus (Johansen) and the other (Bradley) was, for ten years (1996-2006), a secretary of the Society. This version of the NEAS history ends with the forty-ninth symposium held in April, 2010.

Many of the individuals playing prominent roles in NEAS are best known by their nicknames, such as Bea or Bob, and we use these names freely. The full names and affiliations of persons significantly involved in NEAS are given in Appendix D. The names of many others, who played more transitory roles in the symposia, are included in the text, but not in the appendices.

A few notable gaps mar the completeness of the records of symposia held in the 1960s and early 1970s. We would be most pleased to receive information pertaining to symposia in those years.

Abbreviations used in this document are as follows:

NEAS	Northeast Algal Society
IC	Ideas Committee
DC	Development Committee
EC	Executive Committee
PSA	Phycological Society of America
SUNY	State University of New York
MBL	Marine Biological Laboratory
WHOI	Woods Hole Oceanographic Institution

THE EARLY YEARS

In the mid 1960s several phycologists were teaching and carrying out research on algae at various institutions in New York City and the surrounding area (Table 1). Discussions among some of these scientists led to the idea of twice-a-year informal meetings at which they and their graduate students would make presentations and share ideas about their research on algae. The records indicate that from the very beginning a focus of the meetings was on research, both carried out by professional scientists and, most importantly, by students.

The Metropolitan Area Algae Discussion Group. In the spring of 1966, Sheldon Aaronson and Bill Siegelman convened the first meeting of the Metropolitan Area Algae Discussion Group at Queens College in New York City. Aaronson, in a letter dated 26 April 1990, stated that the organization "… was started as a means of expediting communication and cooperation among the people in the metropolitan New York area who were interested in or working with algae." In a document dated October 1969, Bob Wilce stated that the goal of these meetings "…was to provide a vehicle to present and discuss algal research twice a year to the considerable number of phycologists working in the broad metropolitan New York area." According to Aaronson, the attendance and discussions at the first meeting were "surprisingly good."

A second symposium may have been held in the fall of 1966, but no written confirmation of this has come to our attention. However, Norman Lazaroff recalled to Curt Pueschel that this meeting was held at Brookhaven National Laboratory, Upton, NY, and that at least one of the organizers was Bill Siegelman, and possibly also Harvard Lyman (Appendix A). A mailing list dated 21 December 1966 and entitled 'List of persons and institutions interested in informal algal discussion group in the metropolitan New York area' contains the names and addresses of 56 scientists; the compiler is unknown.

Obviously, the name Metropolitan Area Algae Discussion Group was not set in stone, as Ray Jones' announcement of the third meeting was for the "… second [sic] meeting of algal physiologists…" with no mention of the name coined by Aaronson and Siegelman. However, the possibility that the meeting hosted by Jones did not belong to the concept of Aaronson and Siegelman needs to be recognized. We may never know for certain and, unless we hear otherwise, we will assume that the symposium, held in the spring of 1967 at SUNY, Stony Brook, was number three. This third symposium, a one-day meeting, was held on April Fools Day, 1967, and featured two half-hour morning talks, one by Jones on the physiology of *Chlamydomonas* and the other by Leland N. Edmunds on cell division in *Euglena*. After a buffet lunch (for $2.75) seven graduate students gave 15-minute talks about their research, all except one on physiological research on *Chlamydomonas* and *Euglena*.

A fourth symposium was convened on 16 December 1967 at Hunter College, in New York City. It appears that Ruth Sager and Marcia Brody organized this meeting, and they called it the "Fourth Meeting of the Metropolitan Area Algae Discussion Group." According to the program the focus of the Hunter College symposium was, "Studies of Molecular Organization and Organelle Formation with Algae." The four morning papers, all dealing with *Chlamydomonas*, were given by Thomas Cavalier-Smith, Philip Siekevitz, David Ringo, and Ruth Sager. After lunch (for $2), Bill Siegelman, Harvard Lyman and Brody made presentations, two of which concerned *Euglena*. According to Norm Lazaroff, Jerry Schiff (Brandeis University) and two of his post-docs, Lazaroff and Harvard Lyman helped promote these early meetings.

Table 1.

Some of the attendees at the 1966 and 1967 meetings of the Metropolitan Area Algae Discussion Group.

Sheldon Aaronson, Queens College
Melvin M. Belsky, Brooklyn College
Marcia Brody, Hunter College
Leland N. Edmunds, SUNY, Stony Brook
S.H. Hutner, Haskins Laboratories, New York
Raymond F. Jones, SUNY, Stony Brook
Harvard Lyman, Brookhaven National Laboratory
J.J.A. McLaughlin, Fordham University
C.A. Price, Rutgers University
Luigi Provasoli, Haskins Laboratories, New York
Ruth Sager, Hunter College
Melvyn I. Selsky, Brooklyn College
Bill Siegelman, Brookhaven National Laboratory

According to Norm Lazaroff's communication with Curt Pueschel, a fifth symposium was held in the spring of 1968 at Yale University in the so-called 'Tootsie Roll Tower' (*i.e.,* in the Kline Biology Building, with its massive brick-faced columns).

An announcement prepared by conveners Melvyn I. Selsky and Melvin M. Belsky (contrast the names) served to prepare phycologists for the sixth symposium of the Metropolitan Area Algal Discussion Group. It was held on 5 October 1968 at Brooklyn College. There were seven half-hour presentations, all by scientists from New York and vicinity, but the subjects of the talks ranged more widely than on the cell biology of *Chlamydomonas* and *Euglena,* as in the earlier symposia (Table 2). Frank Trainor, then only 39 years old, attended this symposium and volunteered to host the next symposium at the University of Connecticut.

Table 2. *Presentations at the Sixth Metropolitan Area Algal Discussion Group at Brooklyn College in October, 1968.*

Eva K. Hawkins. Cell contact in red algae. The fine structure of pit connections in *Ceramium diaphanum.*
Solomon Goldstein. Isolation and characterization of monocentric chytrids.
Melvyn I. Selsky. Inhibition of photoreactivation of UV-induced injury in *Euglena gracilis* by streptomycin.
David Gimpelson. Primary productivity and metabolism of salt marsh epiphytic communities.
L.N. Edmunds, R.R. Funch, and C.A. Pevney. Effects of high frequency light cycles and 'skeleton' photoperiods on rhythmic cell division and macromolecular synthesis in synchronous cultures of *Euglena.*
J.H. Herz and S. Aaronson. Types of aldolase in microorganisms and its use as an evolutionary marker.
Virginia Bonawit and Marion Himes. Autoradiographic studies of the development of the new hemicell in *Micrasterias thomasainia* – a progress report.

Symposium expansion from New York. The seventh symposium was inspired by Frank Trainor's idea of having the organization become more far-reaching, drawing in scientists from all of New England. This progenitor of NEAS was organized by Trainor and Joanna Z. Page, and was held on 22 March 1969 at the University of Connecticut in Storrs. About 75 persons interested in algae attended the meeting. In their printed announcement, Page and Trainor shortened the name of the organization to '*Algal Discussion Group*' because the participants were from several New England states. The program contained five invited half-hour morning talks and an afternoon symposium entitled 'Development of marine phycology and marine stations.' This was chaired by Bob Wilce, who was ably assisted by Joe Ramus (Yale University) and Art Mathieson (University of New Hampshire). Interestingly, none of the presenters were from the New York metropolitan region, in sharp contrast to previous symposia, and *Chlamydomonas* and *Euglena* were not even listed in the printed program. The spring 1969 symposium was the first involvement of two individuals who would continue to play critical roles in the evolution of NEAS: Trainor and Wilce.

Table 3. *Presentations at the seventh symposium (1969) at the University of Connecticut.*

Carol A. Thompson	*Scenedesmus* cell walls.
Phil Cook	Cytochemistry of host-parasite relationships between *Closterium* and *Ancylistes*.
Mary Schultz	Speciation in the diatom genus *Cyclotella*.
Theodore J. Smayda	The influence of light, temperature and salinity on cell division of the marine diatom, *Detonula confervacea* (Cleve) Gren.
Robert R. Guillard.	Responses of euryhaline and stenohaline diatoms to composition of the medium.

The fall symposium of 1969, number 8, was convened 11-12 October, and there were 135-140 attendees, a surprisingly large number. This meeting was pivotal. It appears that this was the first symposium lasting more than one day, although the first day was used only for registration and an evening "smoker." It was also the first time that the new name appeared: "*Northeast Algal Symposium.*" Later this was easily changed to Northeast Algal Society. This fall symposium was organized by Bob Wilce and his graduate student, Jim Sears, and held at the University of Massachusetts, in Amherst. The week following the symposium, Bob Wilce wrote a summary of the event. Here he noted that some of the attendees came from as far away as Halifax and Quebec City, and some from as far south as Washington, DC. This was a far cry from the much narrower 'Metropolitan Algal Meetings,' first convened in the spring of 1966 at Queens College. This was the first official involvement of Sears, who later also made major contributions to NEAS. The presentations were organized so as to have four half-hour talks and a panel discussion in both the morning and afternoon (Table 4). The morning panelists were Bill Siegleman (Brookhaven National Laboratory), L. Bogorad (Harvard University), and D. Price (Rutgers University). In the afternoon the panelists were Wilce, Walter Adey (Smithsonian Institution), and possibly one other person. It is interesting to note that, forty years later, in 2009, Wilce would again be a symposium convener.

Table 4. *The eight papers presented at symposium number 8 (October, 1969) at the University of Massachusetts.*

Louis A. Hobson	The effect of inorganic nitrogen on macromolecular synthesis by two centric diatoms grown in batch culture.
John Torrey	Experimental studies on *Fucus* embryology.
H. Lyman	Control mechanisms in chloroplast biogenesis in *Euglena*.

R. Jones. Study of differentiation in unicellular algae.

William Woelkerling. Cytology and reproduction in acrochaetoid red algae.

L.A. Hanic. Life history studies in *Urospora* and *Codiolum*.

James R. Sears. Sublittoral algal ecology in southern Cape Cod.

Walter Adey. Coralline algal ecology.

In the fall of 1970 (17 October), Norm Lazaroff hosted the Ninth Northeast Algal Symposium at SUNY, Binghamton. In the one-day meeting, five half-hour talks were given, three of them by scientists from Binghamton (Table 5). None of the presenters were from the New York Metropolitan area, and *Chlamydomonas* and *Euglena* were once again absent from the printed abstracts. An afternoon panel entitled "Concepts of Algal Evolution" featured Sheldon Aaronson (Molecular evidence for algal evolution), Conrad Gebelein (Pre-Cambrian algae), Seymour Hutner (The quintessential flagellate), Steve Golubic (The speciation process in Cyanophyta) and Lynn Margulis (Symbiotic origin of plastids?).

Table 5. *Talks given at the Ninth Northeast Algal Symposium at SUNY, Binghamton, in October, 1970.*

Norman Lazaroff	Cinematographic study of the nostocacean developmental cycle.
Alex Shrift	An induction and maintenance phenomenon in *Chlorella*.
George J. Schumacher	Some epiphytic algae.
John M. Kingsbury	A study by transect of the intertidal vegetation of Star Island, Isles of Shoals.
G. W. Fuhs	Phosphorus-limited growth of plankton algae.

For reasons unknown, we have no records of symposia in the years 1971 and 1972, and possibly none were held. Perhaps it was a lack of impetus from the 1970 symposium, with nobody willing to spend the relatively great amount of time necessary to assemble a gathering. However, it would be folly to upset the sequence, hence, on the assumption that symposia might have been held in 1971 and 1972, we will continue to assign numbers 10 and 11 to those years.

Table 6. *The probable symposia held in the 1960s and in 1970 that represent precursors of NEAS.*

1	Spring 1966	Aaronson, Siegelman	Queens College
2	Fall 1966	Siegelman, Lyman?	Brookhaven
3	Spring 1967	Jones	SUNY, Stony Brook
4	December 1967	Sager, Brody	Hunter College
5	Spring 1968	Unknown	Yale University
6	October 1968	Selsky, Belsky	Brooklyn College
7	March 1969	Trainor, Page	Univ. of Connecticut
8	October 1969	Wilce, Sears	Univ. of Massachusetts
9	October 1970	Lazaroff	SUNY, Binghamton

During these years, in the late 1960s and the early 1970s, Phil Cook (University of Vermont) annually invited several of his phycological friends to spend a weekend at his delightful seaside camp in Maine. Paul Hargraves remembered that it might have been in Ogunquit in a "large white inn on a rocky bluff overlooking the Atlantic." Conversation dealing with the fledgling NEAS was most certainly part of the talk and camaraderie.

On 7 April 1973, Frank Trainor again picked up the ball and convened a spring symposium at the University of Connecticut in Storrs. On the printed program he called it the '*Northeast Algal Discussion Group*,' but did not give it a number, so we will recognize it as the twelfth. The one-day symposium had 21 15-minute talks, many of them oriented towards ecology. At noon of this day, Trainor went to the hospital with an irregular heartbeat, but his students carried on for him in the afternoon. Except for this episode, the meeting was ruled a success, with about 100 attendees.

Again, all the evidence (or lack of) points to no symposium in 1974, the last time a year may have passed without one. However, because of our uncertainty, and for the reasons given earlier, we continue to designate this year as containing symposium number 13.

The fourteenth NEAS symposium was convened by Marilyn Harlin and Roger D. Goos (a mycologist) on 19 April 1975 at the University of Rhode Island. In that single day, they established four sessions, each containing 4-5, 15-minute talks followed by discussions, for a total of 19 presentations. The discussion moderators were (1) L. Provasoli and F. Trainor (Morphogenesis in algae: genetic vs. environmental responses), (2) J. Biggins (Photosynthetic pathways: what more can we learn from algae), (3) Art Mathieson (Seaweed: what is its potential as an applied science?) and (4) Bob Wilce (Biogeography, a dynamic process: what are the 20th century floristic trends in the northeastern coast. Can we explain these?). This marked the last symposium for more than twenty years that was held outside Woods Hole (Appendix A).

Should the symposia continue? It was apparent to many phycologists that an annual symposium was professionally of value to all and also extremely enjoyable. Students of algae and professionals alike could share new concepts, developments in the broad scope of phycology, and, almost magically, a feeling of togetherness and real pleasure, all of which had been characteristic of the symposia held thus far. Therefore, it was not difficult to conjure up reasons for continuing the NEAS symposia. All that was required was an organization and an appropriate venue.

THE WOODS HOLE SYMPOSIA 1976-1997

The next 22 annual NEAS symposia were held at the Marine Biological Laboratory in Woods Hole. This proved to be a very satisfactory setting, with most symposia held in delightful spring weather and the stimulating surroundings of Cape Cod.

The 1970s Woods Hole symposia. The 1976 NEAS symposium, the fifteenth, was hosted by Ray Jones and (again) Bob Wilce. It lasted three days and featured 30, 20-minute papers given on Friday afternoon and all day Saturday, leaving Sunday for ancillary activities. For the first time, Saturday evening was devoted to social activities, including a cocktail hour, banquet, and a distinguished speaker. This format was so successful that it has been propagated, in part at least, to this day. At Woods Hole the Swope Conference Center was admirably suited for these activities. The evening speaker, John Pringle (Halifax), spoke on the "*History and development of sea-*

weed utilization in the Canadian Maritimes." For the first time one of us (Johansen) heard the enchanting Moss-ing Ballad, a haunting song commemorating the harvesting of Irish moss (*Chondrus crispus*) in the Canadian Maritimes. On Sunday, attendees were invited to tour John Ryther's macroalgae culture laboratory near Woods Hole Oceanographic Institution (WHOI), and then some of the attendees joined a seaweed collecting trip to the in-tertidal (Nobska jetty).

The sixteenth NEAS symposium, in 1977 (29-30 April), was hosted by three co-conveners, Joe Ramus, Sam Beale and (again) Frank Trainor, who ran an ambitious meeting featuring 30 talks. They were organized into three half-day sessions, the first on Friday afternoon and the other two on Saturday. In the circular, the conveners used a slew of adjectives: "sunny spring weekend, truly illuminating papers, scintillating informal discussions, ex-cellent cuisine and informative and enjoyable field trips." The social hour and banquet were on Friday evening. A highlight of the meeting was the distinguished speaker, the irrepressible George F. Papenfuss, who spoke on "*Landmarks in the discovery of sexuality and alternation of generations in brown algae*" after the banquet. Pa-penfuss, emeritus professor from the University of California at Berkeley, was known internationally not only for his scientific papers, but also for his enthusiastic way of helping fledgling phycologists. As a former Papenfuss student, I (Johansen) consider him a marvelous human being with very few equals. Added features were Craig Schneider and Charlie Yarish leading a field trip to the intertidal and Jim Sears preparing a "delectable Irish moss blanc mange."

The 1978 (28-29 April) NEAS symposium, the seventeenth, was another exciting one featuring a world-renowned phycologist, Wm. Randolph Taylor. Taylor, emeritus professor from the University of Michigan, gave a talk on Saturday afternoon entitled "*Early appearance of northeastern algae in American floras and textbooks, leading to the algal studies developed at Woods Hole.*" The symposium, hosted by Steve Golubic and Bill Jo-hansen, was held on Friday and Saturday. It had the usual happy hour and a banquet highlighted by a humorous banquet address by Frank Trainor entitled "*How do your algae grow?*" In his talk, Trainor went over the skills re-quired to concoct satisfactory media and then presented imaginative and humorous phycological distortions with illustrations, for example, twisting *Scenedesmus* into "skinny-dead-mouse." The Gray Museum and the MBL Herbarium were open for inspection courtesy of Wesley N. Tiffany and Edwin Moul.

The last NEAS symposium of the 1970s (27-28 April 1979), the eighteenth one, was hosted by Annette Cole-man and Jim Sears, two scientists who subsequently played important roles in the development of NEAS. A first-time event this year was the presentation of posters, and there were seven. Thirty 15-minute talks highlighted the symposium. Following the Friday evening banquet, distinguished speaker John D. Dodge (Royal Hollaway Col-lege, Surrey, England) gave a superb talk entitled "*Dinoflagellates: animal, vegetable and mineral.*" Wesley Tiffany and Ed Moul of the Gray Museum and MBL Herbarium set up displays of Wm. Randolph Taylor's collec-tions of algae from the Dry Tortugas. Cultures of Annette Coleman's freshwater algae were also on display.

Woods Hole in the 1980s. In 1980 (2-3 May) the nineteenth NEAS symposium convened under the leadership of Phil Cook and Craig Schneider, the latter destined to play a prominent role in the development of the society. A new wrinkle this year were three special half-hour invited lectures labeled a 'minisymposium' on Saturday morn-ing: Malcolm Brown (University of North Carolina), on "*Biosynthesis of cellulose, nature's most abundant macromolecule,*" John van der Meer (Atlantic Research Laboratory, Halifax), on "*Use of genetics in phycologial research,*" and Jeremy Pickett-Heaps (University of Colorado), on "*Cell division in diatoms.*" The distinguished after-banquet speaker was F.J.R. "Max" Taylor (University of British Columbia) whose title was "*The origin of* *eukaryotic cells.*" An added feature this year was a petition to MBL to restore the marine botany course that was among the first courses offered at that institution and a regular summer offering for many years. Co-conveners Cook and Schneider, recognizing the disappointment felt at the elimination of this course, penned a letter to Dr. Paul Gross, the Director of MBL, and obtained 81 signatures from symposium attendees. In the letter they asked that the marine botany course be "re-established" in the near future. Gross' reply, dated 24 June 1980, stated that the course had merely been suspended for one or two years in order for a committee to review its feasibility. He stressed that extramural funds must be found to support it. As far we know, the MBL marine botany course has never again been taught since the review.

The twentieth NEAS symposium was held 11-12 April 1981, and hosted by Nina Allen and Art Mathieson. This meeting, moved to Saturday and Sunday (previously they were on Friday and Saturday), contained many events. Lynn Margulis (Boston University) gave an after-banquet keynote address entitled *"Evolution of algae.*" There were four special lectures in a minisymposium on plant-animal interactions: David A. Schoenberg (Medical University of South Carolina), on *"Symbiosis between microalgae and invertebrates,*" Bob Vadas, (University of Maine), on *"Comparative algal-sea urchin interactions in boreal and tropical waters*", Joy A. Geiselman, (University of Montana), on *"Ecology of chemical defenses of algae against the herbivorous snail,* Littorina littorea, *in the New England rocky intertidal community,*" and Don Cheney, (Northeastern University), on *"Plant-gastropod interactions in the intertidal zone: a critique of current concepts.*" At the banquet, Bob Wilce gave a brief talk on the history of NEAS. Aimlee Laderman's contributions to the symposium were an exhibit of the work on diatoms by P.S. Conger and a post-symposium field trip to the well-known Atlantic White Cedar Swamp. As before, the algal herbarium at MBL was open, thanks to Ed Moul. On Sunday Gerald T. Boalch (The Laboratory, Citadel Hill, Plymouth, UK), spoke on *"A history of British phycology.*" Still, like other NEAS symposia, the 15-minute-long contributed papers and the posters were the heart of the meeting.

The twenty-first symposium in 1982 (1-2 May) was pivotal for NEAS, because, as explained later, the members present at a business meeting held during the symposium authorized for the first time the establishment of an Executive Committee. This action was generated by an ad hoc committee, as will be explained later. Thus, the hosts of the 1982 meeting were the last to be selected by the arbitrary way hitherto used. Nevertheless, the 1982 symposium was well-organized by three scientists: Beth Gantt, Hank Parker, and Charlie Yarish. The distinguished speaker was Johan Hellebust (Toronto University), who spoke following the banquet on "Osmoregulation and salt tolerance of algae." On Sunday afternoon a minisymposium entitled "Implication of genetic relationships from algal studies at the micro and macro level" was convened. The invited speakers were Annette Coleman (Brown University), on *"Diversity of chloroplast DNA types among algal phyla,*" Gary L. Floyd (Ohio State University, with Charles J. O'Kelly) on *"Phylogeny of the Ulvophyceae: an ultrastructural perspective and update,*" John van der Meer (Atlantic Research Laboratory, Halifax), on *"Zygote amplification: a recurrent but not universal theme in the Rhodophyta,*" and Michael Neushul (University of California, Santa Barbara), on *"The evolutionary implications of kelp hybridization.*"

The 1983 NEAS symposium, the twenty-second (7-8 May), was organized by Bud Brinkhuis and Peter Heywood and had almost 200 attendees. The program (Figure 1) included 59 abstracts, 31 for contributed papers, 24 for posters, and 4 invited lectures. So many abstracts were submitted that, unfortunately, some had to be rejected. The distinguished speaker was Lynda Goff (University of California at Santa Cruz), on *"Red algal parasites: what are they trying to tell us?*" A minisymposium was held on Sunday afternoon; it was entitled "Seaweed/algal mariculture." Invited speakers were Art Mathieson (University of New Hampshire) who provided an overview on sea-

Figure 1. Logo for the 22nd symposium (1983). Photographs of seaweed farms in China provided by X. G. Fei.

weed mariculture, R.G.S. Bidwell (Atlantic Research Associates Ltd.) who discussed the onshore cultivation of *Chondrus crispus* in Nova Scotia, Ami Ben Amotz (the Israel Oceanographic and Limnological Research, and on sabbatical at Georgia Institute of Technology), who summarized the status of glycerol and Beta-carotene production from *Dunaliella* and Dennis Hanisak (Harbor Branch Oceanographic Institute, Fort Pierce, FL) who reviewed progress on the mariculture of *Gracilaria* in Florida. At this symposium, awards were presented to graduate students for the best talk and the best poster for the first time. The winners were Al Steinman (University of Rhode Island) and John Hackney (Georgetown University).

In 1984, Llewellya Hillis-Colinvaux and Don Cheney convened the twenty-third NEAS symposium on 28-29 April (Fig.2). This was the first meeting of the NEAS as a section of the Phycological Society of America. From now on the official name would be Northeast Algal Society, not the Northeast Algal Symposium. This was the biggest NEAS meeting so far, being attended by 200 professional scientists and graduate students from many parts of North America, as well as from Great Britain, Argentina and Brazil. There were 53 oral presentations and 20 posters, a fact that required concurrent sessions on Saturday (for the first time). There were several highlights. Wm. Randolph Taylor was presented an endowed chair in MBL's Lillie Auditorium at the Saturday evening banquet where Wilce pointed out some humorous episodes in Taylor's long career. Donations of more than $1000 resulted in a brass engraved plate affixed to Taylor's Chair in Lillie Auditorium. This year's distinguished speaker, Holger Jannasch (WHOI), presented a provocative lecture on *"Plant life in the deep sea."* Awards went to graduate students Donna Johnson (University of Connecticut) and Rod Fujita (MBL) for their stellar presentations. A minisymposium featured four talks on adaptations of algae to physical and biological factors. The speakers were Joe Ramus (Yale University), on *"Cellular and structural enhancement of light capture,"* Penny Chisholm and co-workers (MIT), on *"The regulation of cell cycle progression in marine phytoplankton in periodic environments,"* Rachel Ann Merz (Northeastern University), on *"Benthic organisms and hydromechanical adaptations,"* and Nancy Targett (Skidaway Institute of

Figure 2. William Randolph Taylor, as he appears on the cover of the abstract booklet for symposium number 23 (1984). Cover designed by Sue Rieter.

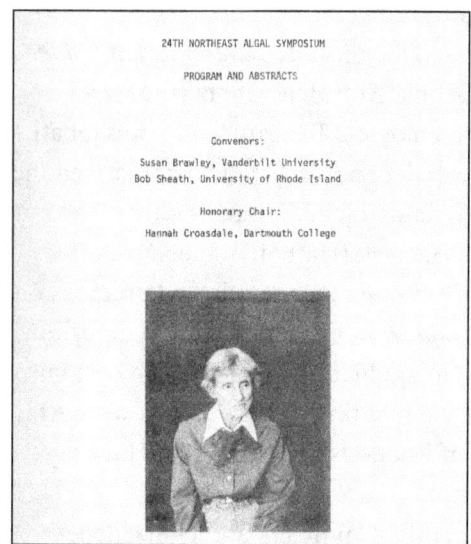

24TH NORTHEAST ALGAL SYMPOSIUM

PROGRAM AND ABSTRACTS

Convenors:
Susan Brawley, Vanderbilt University
Bob Sheath, University of Rhode Island

Honorary Chair:
Hannah Croasdale, Dartmouth College

Figure 3. Hannah Croasdale, as she appears in a photograph inside the cover of the abstract booklet for symposium number 24 (1985).

Oceanography, University of Georgia), on "*Chemical-biological interactions in the marine environment: chemical defenses to herbivory.*"

The twenty-fourth NEAS symposium, 27-28 April 1985, was convened by Susan Brawley and Bob Sheath, with more than 200 attendees. Thirty three papers were presented and 14 posters prepared. Hannah Croasdale (Fig. 3) (Dartmouth College) was selected as honorary chair for her long persevering work on desmids, and for her indomitable spirit. The NEAS logo this year appropriately showed a dividing desmid, *Xanthidium antilopaeum* (Fig. 4). At the banquet, Hannah took questions from the floor, answering them with a quick wit and at some length. The graduate student awards went to Sara Lewis (Harvard University) for her talk and Marth Ludwig (McGill University) for her poster. The banquet speaker was Ursula Goodenough (Washington University) whose lecture was entitled "*Evolutionary relationships between the sexual agglutinin and the cell wall of* Chlamydomonas." A Sunday morning symposium dealt with "*Acid rain in the Northeast.*" The speakers were Chris Bernabo (Director of the Interagency Task Force on Acid Precipitation), on "*The U.S.*

National Acid Precipitation Assessment Program," Paul Godfrey (University of Massachusetts), on "*Seasonal and geographic variations in pH and alkalinity of Massachusetts surface water,*" Ken Nicholls (Ontario Ministry of the Environment), on "*Phytoplankton and lake acidification—A review with emphasis on experiences in Ontario, 1973-1983,*" Ron Davis (University of Maine), on "*Effects of changes in water chemistry inferred from algal remains in sediments,*" and Ann E. Umbach (Cornell University), on "*Cellular mechanisms of acid tolerance by freshwater algae.*" Three workshops were given on Sunday afternoon: Valerie Vreeland (University of California, Berkeley), on "*Monoclonal antibody production strategies,*" Lionel F. Jaffe (MBL), on "*The vibrating probe,*" and John Walrond (MBL), on "*Rapid freezing.*"

By 1986, NEAS symposia were scientifically and socially exciting and attracting phycologists from far and wide. In that year (19-20 April 1986), NEAS celebrated its twenty-fifth anniversary with a most successful symposium run by co-conveners Bea Robinson and Larry Liddle. More than 160 persons attended, and 24 papers were

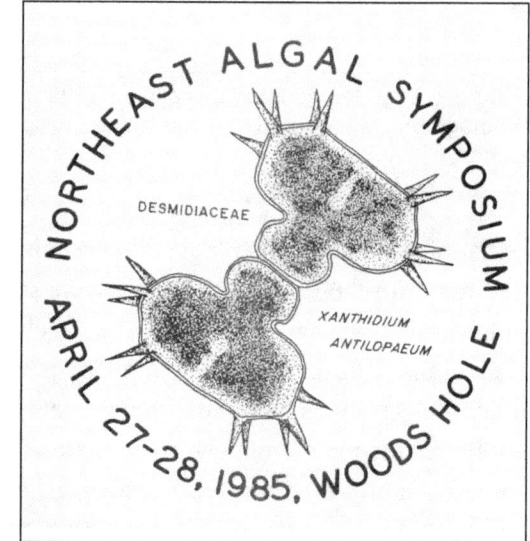

Figure 4. Logo for the 24th symposium (1985). A recently divided desmid, *Xanthidium antilopaeum.* Contributed by Glen Thursby.

presented and 12 posters were prepared. A highlight of the 1986 meeting was that Luigi Provasoli was honorary chair (Figure 5). After coming to the United States from Italy, Luigi spent much of his scientific life at Haskins Laboratories where he did beautiful culture experiments on unialgal protists. His life (and that of his wife Rose) was summarized in a humorous mock interrogation by Bob Wilce and Bill Johansen at a banquet roasting. The logo for this meeting consisted of a montage of four drawings of *Euglena* designed by Deirdre O'Connor, a

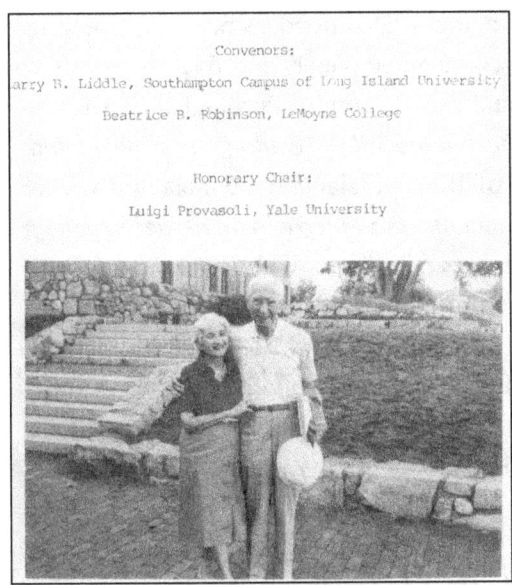

Figure 5. Luigi and Rose Provasoli as they appear together on the inside cover of the abstract booklet for symposium number 25 (1986).

student at Le Moyne College (Figure 6). *Euglena* was an organism of much interest to Provasoli.

Juan Correa (Acadia University) won the award for the best presentation. The banquet speaker was Mike Levandowsky, indirectly a colleague of Provasoli from Haskins Laboratories, Pace University, who spoke on "*Signals and signal processing in the algal protists.*" Two minisymposia were organized. Minisymposium I, entitled "*Phycology at WHOI*," featured researchers from the Institute: Don Anderson on "*The biology and ecology of toxic dinoflagellates in New England*," Joel C. Goldman and D. A. Caron on "*Dynamics of phytoplankton-protozoan interactions in homogeneous and heterogeneous environments*," R. J. Olson and co-workers on "*Flow cytometric studies of* Synechococcus *distributions and properties*," Diane K. Stoecker on "*Chloroplast retention by marine planktonic ciliates: ecological and evolutionary implications*," and John B. Waterbury on "N_2 *fixation in tropical unicellular cyanobacteria.*" Minisymposium II, "*What's new in kelp biology*," had talks by Valerie A. Gerard (SUNY at Stony Brook), on "*Searching for light ecotypes of* Laminaria," Tom A. Dean (University of California, Encinatas), on "*Recruitment of giant kelp: effects of physicochemical factors on early life-stages*," Stephen R. Fain (Simon Fraser University), on "*Molecular evolution in the Laminariales: restriction analysis of chloroplast DNA*," and Robert E. Scheibling (Dalhousie University), on "*Macroalgal succession following mass mortalities of sea urchins off Nova Scotia.*" The first NEAS group photo was taken (Figure 7). A suggested permanent logo for the Society was selected at the all-symposium business meeting (see Fig. 29, page 36). On the warm and sunny Sunday afternoon, Aimlee Laderman led a field trip to the Atlantic White Cedar Bog.

In 1987 (25-26 April), Paul Hargraves and Tom Lee convened the twenty-sixth symposium. A stimulating program of 32 paper presentations and 13 posters, many by graduate students, highlighted the gathering. It was attended by about 135 persons. The award for the best talk by a graduate student was won by Rick Greene (SUNY, Stony Brook). The keynote speaker, Matthew Dring (University of Belfast), spoke on "*Throwing light on seaweeds.*" This was the first time the distinguished speaker gave his/her lecture before the banquet, rather than afterward. It was a lovely mix of science and humor on light, pigments and photosynthesis in marine algae. For the second straight year a group photo of the attendees was taken (Figure 8). The 1987 logo was a drawing of *Codium* produced by the NEAS Development Committee (Figure 9). Both an honorary chair and minisymposia were absent from the 1987 symposium. Incidentally, this was the first year that onsite computers were used at the registration table.

Figure 6. Logo for the 25th symposium (1986), showing *Euglena gracilis*, designed by Deirdre O'Connor.

The twenty-seventh NEAS symposium met on 22-24 April 1988, with about 100 attendees present. Co-conveners Phil Sze and Bob Vadas put together a stimulating program of 25 contributed papers and five posters. Two keynote speakers were featured: Bob Paine (University of Washington) and Carole Lembi (Purdue University). Paine spoke on "*A zoologists perspective on marine benthic algae as superb material for ecological experimentation,*" including references to coralline algae on the wave-exposed rocks of Tatoosh Island at the most northwestern point of the United States. Lembi's talk on "*Static freshwater environments and macrophytic algae: are they compatible?*" was a discussion of the roles of freshwater filamentous algae, and her work with *Pithophora*. The 1988 honorary chairperson was to have been Ed Moul; sadly it was in memoriam. Moul, age 86, passed away 16 April, just one week before the symposium. We were pleased, however, that Ed's widow, Mrs. Hazel Moul and her close friend Louise Bush, joined the banquet. Following the meal John Kingsbury (Cornell University) and Frank Trainor reminisced about Moul. For the first time this year the DC had enough money in their coffers to make a travel award to a student; $200 went to Juan Correa (Atlantic Research Laboratory, Halifax). As in several previous years, Aimlee Laderman led a post-symposium trip to the local White Cedar Bog. The logo for the symposium included a species of *Alaria* and one of *Gonium*, designed by Phil Sze (Figure 10).

In 1989 (29-30 April), Bob Steneck and John van der Meer convened the twenty-eighth NEAS symposium, drawing 113 attendees. R. C. Sokol (SUNY at Albany) won the graduate student award for the best presentation. Lois Pfiester (University of Oklahoma), the distinguished speaker, gave a delightful pre-banquet talk entitled "*The study of freshwater dinoflagellates: luck and elbow grease.*" The logo for this symposium was a drawing of a nodule of *Lithothamnion glaciale* rendered by Kirt Moody (Figure 11).

Figure 7. A group photograph of attendees standing in front of Lillie at the 25th symposium (1986).

Woods Hole in the 1990s. Eight symposia were held at the Marine Biological Laboratory (MBL) in Woods Hole in the 1990s until, in 1998, circumstances necessitated a change to a new venue. NEAS was now on a roll and each symposium was highly successful both scientifically and socially.

In 1990 (27-29 April), Ruth Schmitter and Pete Siver convened the twenty-ninth NEAS symposium (Figure 12) with 100 registrants, 18 papers, and 12 posters. The late Bud Brinkhuis (SUNY at Stony Brook) was the honorary chair. Bud would have begun his two-year term as chairperson of NEAS had he not passed away suddenly in the summer of 1989. Charlie Yarish gave an oral tribute to Bud at the Saturday banquet The distinguished speaker was Bruce Parker (Virginia Polytechnic Institute and State University), a renowned science historian. Parker spoke on "*Origins, evolution and development of phycology in North America.*" This year's minisymposium was on "*Algae and human affairs.*" The speakers were Ric Breese (Manville Sales Corporation), on "*Diatomite—the uses of fossil algae,*" Karen Steidinger (Florida Department of Natural Resources), on "*Marine dinoflagellate blooms: dynamics and impacts,*" and Dimitri Stancioff and Gordon Guist (both from Marine Colloids), on "*Old and new uses of carrageenan.*" A poster honoring the memory of Barry Egan (Patrick F. Egan) was prepared by Frank Trainor and Charlie Yarish; Barry had died early in 1990 after earning his doctorate in 1988 at the University of Connecticut. Barry was an enthusiastic phycological researcher with an astoundingly

Figure 8. A group photograph of attendees at the 26th symposium (1987).

large number of publications to his credit, many of them co-authored with Trainor and Yarish. This was the first year that the student awards for the best presentations were named the "Robert T. Wilce Awards." The honor implied by this came as a complete surprise to Wilce. The first Wilce award winners were Janet E. Kuebler (University of Maine) and Josh P. Philibert (University of Massachusetts). By 1990, commercial firms were regular exhibitors at the symposium and this year they were: Connecticut Valley Biological Supply, Manville Sales Corp., Marine Colloids, Don Santo

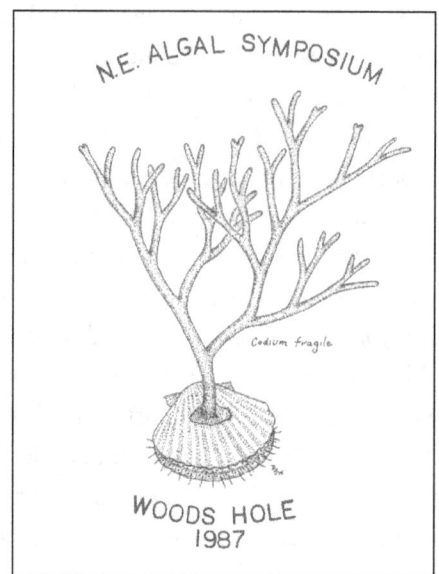

Figure 9. Logo for the 26th symposium (1987). Codium fragile on a scallop. Artist unknown.

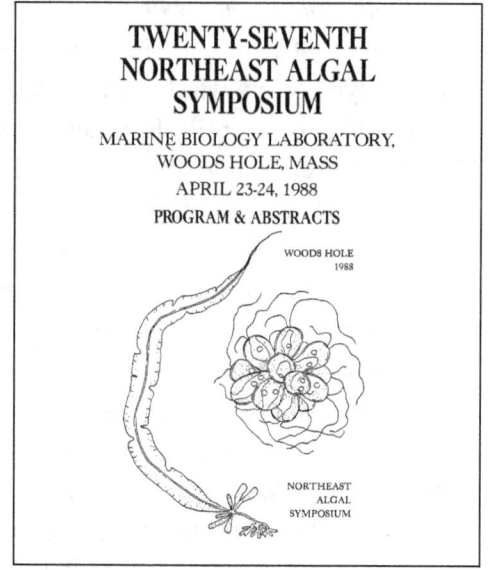

Figure 10. Logo for the 27th symposium (1988). A species of Alaria and one of Gonium. Designed by Phil Sze.

Microscope Corp. (Nikon), Microtech Optical (Olympus), Carl Zeiss, Lubrecht and Cramer (Scientific books), and Solar Components. The logo, designed by Matt Kornhaas, featured *Mallomonas* and *Ectocarpus* (Figure 12).

The thirtieth NEAS symposium occurred at the MBL in Woods Hole on 27-28 April 1991. It was hosted by Dick Fralick and Jim Sears. With 115 registrants, 24 oral presentations, 16 posters, two distinguished speakers, and two minisymposia, as well as the other symposium events, the schedule was full. The Wilce award winners were Julie Yates (Smith College) and Katherine Duff (Queen's University). The two distinguished speakers had both played essential roles in the development of NEAS through the years: Annette Coleman and Bob Wilce. Coleman's lecture was on "*Species concepts in algae: contributions from molecular biology*" and Wilce's was on the "*Role of the Arctic Ocean as a bridge between the Atlantic and Pacific Oceans.*" Wilce was also the honorary chairperson and received from the Society a document entitling him to lifetime membership in NEAS. During the banquet several of his former students and colleagues showed interesting and humorous slides of his work in the Arctic and sub-Arctic. The admiration and respect for Wilce's research, as well as for his efforts on behalf of NEAS, were very evident. The first minisymposium was entitled "*Algal phytogeography*" and featured presentations by Craig Schneider (with Richard B. Searles) on "*Biogeography of seaweeds in the warm temperate Carolina region*" and Max Hommersand (with Suzanne Fredericq; University of North Carolina), on "*Biogeographic affinities of the red algae of the temperate North Atlantic Ocean.*" The second minisymposium, "*Contributions from molecular biology to phy-*

Figure 11. Logo for the 28th symposium in 1989, *Lithothamnion glaciale.* Drawn by Kirt Moody.

cology," had talks by Debashish Bhayttacharya (MBL), on "*Molecular phylogenetic analysis of algae using actin coding regions*" and Linda Goff (University of California in Santa Cruz), on "*The ribosomal DNA internal transcribed spacer regions are conserved at the species level in the red alga* Gracilariopsis lemaneiformis." The DC gave out several travel and book awards to students. Two commercial firms attending also gave out awards based on random drawings: Connecticut Valley Biological Supply awarded a $50 prize and Swift Microscopes a handsome barometer. The logo featured *Gonium* and *Laminaria solidungula* (Figure 13).

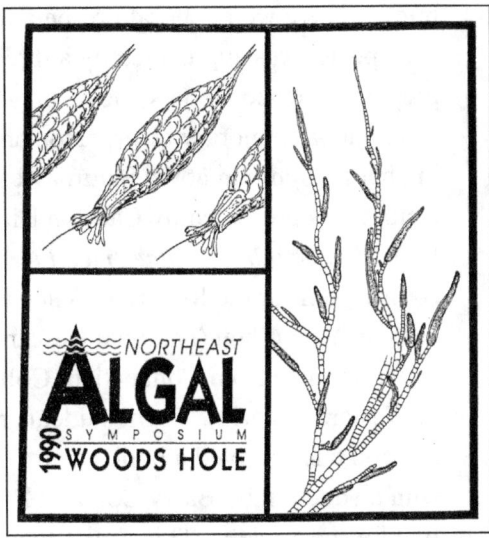

Figure 12. Logo for the 29th symposium (1990). *Mallomonas* and *Ectocarpus*. Designed by Matt Kornhaas.

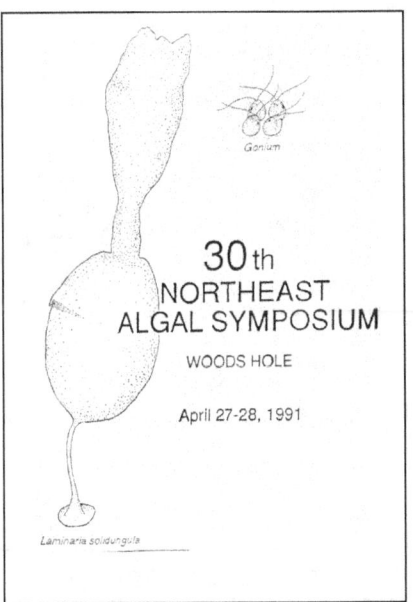

Figure 13. Logo for the 30th symposium (1991). *Laminaria solidungula* and *Gonium*. Artist unknown.

In 1992 (24-26 April), Aimlee Laderman and Mike Levandowsky hosted the thirty-first NEAS symposium at the MBL in weather that was unusually subpar. With 125 registrants, 20 contributed papers, and 25 posters, two distinguished lectures, an honorary chairperson, and a minisymposium, the symposium was a busy one. Two graduate students were honored for their presentations by winning the Wilce awards: Steve Wilhelm (University of Western Ontario) and Kimberly Brown (Queen's University). The distinguished lecturers were Frank Trainor who spoke on "Scenedesmus *phenotypic plasticity: cyclomorphosis*" and Don Anderson (Woods Hole Oceanographic Institute), who spoke on "*Toxic dinoflagellate blooms and red tides in New England and abroad: a biogeographic and physiological perspective.*" Trainor was also the honorary chairperson. A minisymposium on problem algae featured speakers A.W. White (National Marine Fisheries Service), on "*Fate and consequences of paralytic shellfish toxins in the marine food web,*" Tracey Villareal (University of Texas), on "*Domoic acid and amnesiac shellfish poisoning: a review,*" and Don Cheney (Northeastern University), on "*Project development to harvest the beach-fouling alga* Pilayella littoralis *in Nahant Bay, MA.*" The logo, drawn by Levandowsky aided by Laderman, shows life cycles of *Scenedesmus* and *Alexandrium* (Figure 14). About half of all registrants were students, with nine of them winning DC awards of travel money or books.

Barry Colt and Hank Parker convened the thirty-second NEAS symposium at the MBL in 1993 (24-25 April) with 133 regis-

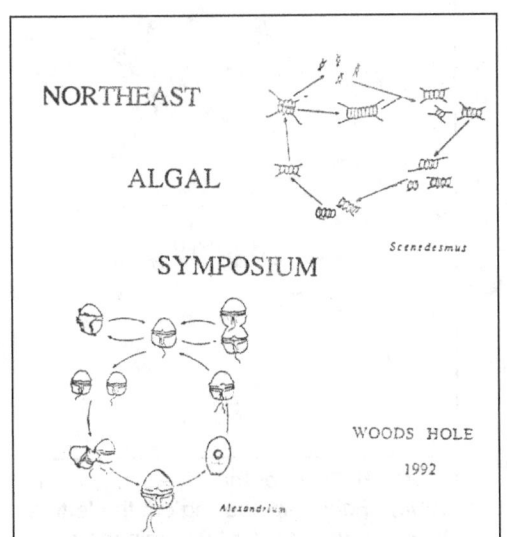

Figure 14. Logo for the 31st symposium (1992). Life cycles of a *Scenedesmus* sp. and an *Alexandrium* sp.

Figure 15. Logo for the 32nd symposium (1993). A *Gracilaria* sp. and a *Micrasterias* sp. Drawn by Barry Colt.

trants. There were 23 oral presentations and 17 posters. For the first time each poster was augmented by a three-minute oral presentation by the preparer. The conveners chose long-time supporter of NEAS, Annette Coleman, as honorary chairperson and she enthusiastically responded during her recognition at the Saturday evening banquet. The distinguished lecture was presented by Carolyn Bird (National Research Council, Halifax): "*Molecular and other meanderings among the Rhodophyta.*" The competition for the Robert T. Wilce awards was so close that three were chosen: Y. H. Zhou (National Research Council), Anne-Marie Lott (an undergraduate from Connecticut College), and Kelly Murphy (University of Western Ontario). Eight students received $50 travel awards from the DC and one received a book award. A post-banquet social hour combined with an auction by Barry Colt and Hank Parker was great fun and lucrative for the Society. This year's logo incorporated a two-piece drawing of *Gracilaria* and *Micrasterias* by Barry Colt (Figure 15).

The thirty-third NEAS symposium, in 1994 (23-24 April), convened by Paulette Peckol and Curt Pueshel, drew 99 registrants of which almost half were students. Twenty talks, 11 posters, and two distinguished lectures highlighted the event. The Wilce awards were won by Andrea Nerozzi (Brown University) and Christopher Harley (also from Brown University). Linda Graham (University of Wisconsin) presented the distinguished lecture entitled "*Ecological and evolutionary importance of dissolved organic carbon utilization by charophycean algae.*" James T. Carlton (Director, Williams-Mystic) presented the other distinguished lecture on "*Botanical roulette: the potential role of ballast water in the introduction of algal species to North America.*" DC awards for travel money and senior scholarships went to 13 students. Also incorporated into the program was a post-banquet auction run by Barry Colt and Tom Lee. The logo, an artful drawing of four kelp spelling NEAS (Figure 16), was designed by Curt Pueschel and drawn by Nancy Haver.

In 1995 (29-30 April), Dorothy Swift and Gary Wikfors convened the thirty-fourth NEAS symposium at the MBL. Besides the contributed papers and posters, Swift and Wikfors assembled a symposium with a distinguished speaker who was also the honorary chairperson and a Sunday minisymposium with seven speakers. The honorary chairperson and distinguished speaker, Bob Guillard (Bigelow Laboratory for Ocean Sciences), made his presentation on "*Natural history and the marine phytoplankton: some directions taken, seen through a case history.*" The minisymposium, entitled "*Algae in contemporary environmental issues— regulator and regulated*" featured Robert W. Kortmann (Ecosystem Consulting Service, Coventry, Connecticut), on "*Physical, chemical, and biological algae control strategies for eutrophic lakes*", Ted Smayda (University of Rhode Island), on "*Microalgal blooms: human influence*

Figure 16. Logo for the 33rd symposium (1994). Four kelp spelling out the letters NEAS: *Alaria esculenta, Laminaria digitata, Agarum clathratum, Laminaria saccharina*. Designed by Curt Pueschul and drawn by Nancy Haver.

and the open niche hypothesis," Don Cheney on "*Macroalgal blooms: environmental causes and economic opportunities*," Paul G. Falkowski (Brookhaven National Laboratory), on "*Molecular ecology of iron limitation in phytoplankton photosynthesis*," Glen Thursby on "*Macroalgae and toxicity testing: methods development and limitations*," James Foertch, John Swenarton, and Milan Keser (all from Northeast Utilities, Millstone Nuclear Power Station, Waterford, Connecticut), on "*Algae and environmental monitoring: perspectives of the 'regulated'*," and Lynda Goff and co-authors on "*Applications of molecular research to macroalgal ecological studies.*" There were also several ancillary events, such as an historical walking tour of Woods Hole, a tour of the MBL Library, a NEAS auction run by Barry Colt, a post-symposium tour of the MBL Marine Resources Center, and a bicycle ride with Greg Boyer. The logo was a drawing of a set of four diatom species spelling NEAS (Figure 17). Unfortunately, data on the Wilce award winners are lacking.

Figure17. Logo for the 34th symposium (1995). Four taxa of diatoms spelling NEAS: *Thalassionema nitzschioides, Asterionella japonica, Biddulphia alterans, Rhizosolenia robusta.*

The thirty-fifth NEAS symposium was held in 1996 (27-28 April) under the directorship of Carolyn Bird and Greg Boyer. It was attended by 159 participants, 68 of whom were students. The scientific program featured 16 contributed oral presentations and 28 posters. The Wilce awards were won by Keith Josef (Syracuse University) and Susan J. Babuka (SUNY at Binghamton). Before the social hour on Saturday, the distinguished lecture was presented by Bob Anderson (Bigelow Laboratory for Ocean Sciences): "*Algal biodiversity and its significance.*" The theme of algal biodiversity was aptly displayed by the logo, a phylogenetic tree of the algae based on SSU rDNA sequences designed by Bird (Figure 18). At the pre-banquet social hour kelp and dulse canapés were supplied with great flair by Carl Karush of Maine Coast Sea Vegetables. A Sunday mini-symposium on "*Brown tides, a new wave of harmful algal blooms*" featured three talks: Robert Nuzzi (the Suffolk County Department of Health Services) on "*The brown tide in eastern Long Island: thoughts on possible causes based on field studies*," Gregory A. Tracey (Science Applications International Corp., Narragansett, RI) on "*Physiological effects of brown tide on bivalve populations in New England waters*," and Tracy Villareal (University of Texas) on "*The Texas brown tide: ecology of a 6 year bloom.*" A discussion session entitled "*The value of phycological research for industry and government*" followed these talks. Participants were Steve Crawford (Coastal Plantations International), on "*Seaweed aquaculture*," Chris Heinig (MER Assessment Corp.), on "*Microalgal feeds in aquaculture*," Paul Behrens (Martek Biosciences Corp.), on "*Value-added algal products*," Robert Nuzzi (the Suffolk County Department of Health Services, Riverhead, NY), on "*Harmful algal blooms*," Michael Connor (Massachusetts Water Resources Authority), on "*Implications of coastal and freshwater*

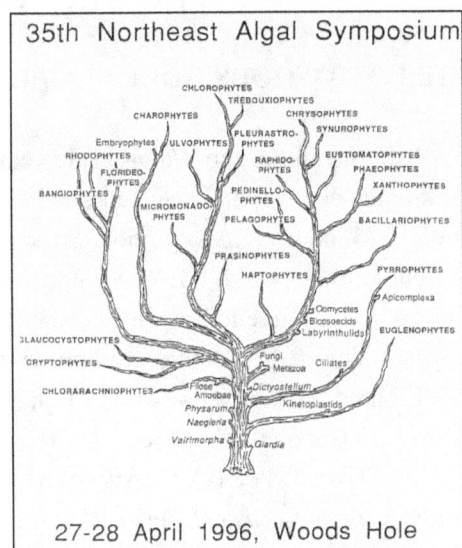

Fig. 18. Logo for the 35th symposium (1996). A phylogenetic tree emphasizing algae based on SSU rRNA sequences in the style of Ernst Haeckel. Drawn by Carolyn Bird.

pollution," and Andrew Bertocci (Algae Tech Seaweed Solutions), on "*Advanced communications*." For the first time a photograph competition was held in both micrograph and macrograph categories. The micrograph winner was Brian Wysor (Southampton College, Long Island University), with Paul Cancellieri (also from Southampton College) second. In the macrophyte competition, both first and second places went to students from the University of Maine: Ester Sarrao took first and Li Rui second. Another first in 1996 was the Frank Shipley Collins Award for meritorious service to the NEAS and to phycology; Ed Boger was the very deserving recipient (Table 12). With his effervescent personality, Ed was a mainstay of the NEAS Development Committee ever since it was established in 1984. On Sunday afternoon, Aimlee Laderman led a tour of the local cedar wetlands.

In 1997 (26-27 April), Glen Thursby and John Wehr convened the thirty-sixth NEAS symposium, the last to be held at Woods Hole until the present one in 2011, a hiatus of fourteen years. The symposium included pre-banquet distinguished speaker Bob Sheath (University of Guelph), who spoke on "*Freshwater red algae: from the molecule to the globe*." Also on the schedule was a minisymposium entitled "*Why should we care about molecular biology?*" featuring two speakers; Barbara MacGregor (Northwestern University), who spoke on "*Applications of molecular methods in aquatic environments*" and Lynda Goff (University of California, Santa Cruz), who spoke on "*Algal parasites, plasmids and viruses: what molecular analyses have revealed*." The Frank Shipley Collins Award was given to Barry Colt (Southeastern Massachusetts University), who was for many years the backbone of the DC. Barry personified the spirit of helpfulness that has been so important in the progress of NEAS through the years. After his passing in 2006, the DC would be named for him. The logo was a drawing of seven common seaweeds (Figure 19).

Figure 19. Logo for the 36th symposium (1997). Seven common New England seaweeds. Artist unknown.

THE POST-WOODS HOLE SYMPOSIA 1998-2010

With the end of the Woods Hole years the EC looked into various conference centers, and selected one in Plymouth, Massachusetts, for the next symposium. This was the first year since 1976 that a NEAS symposium was held outside Woods Hole.

The thirty-seventh NEAS Symposium was held 3-5 April 1998, at the Sheraton Inn in Plymouth, Massachusetts. The co-convenors were Tracy Villareal and Joby Chesnick. There were 104 participants, and the scientific program included 17 contributed oral presentations, 26 posters, and three invited lectures. Three graduate students shared the Wilce awards: Douglas McNaught (University of Maine), Todd Harper (University of New Brunswick), and Bradley Metz (Northeastern University). Saturday's sessions were followed by the distinguished speaker Mike Wynne (University of Michigan), who spoke on the topical subject: "*The complexity of algal systematics: the organisms! the jobs! the future!*" Sunday's program included two invited lecturers: Michael Kuchka (Lehigh University), on "*Chloroplast gene expression in* Chlamydomonas *– what's the nucleus got to do with it?*" and JoAnn M. Burkholder (North Carolina State University), on "*The story of* Pfiesteria*: lessons in biology, toxicology, policy, and scientific ethics*." The meeting logo was a representation of the complicated life cycle of *Pfiesteria* (Figure 20). Interestingly, the generic name *Pfiesteria*, coined in 1996, memorializes Lois Pfiester who was the NEAS distinguished speaker at the 28th symposium in 1989. The Frank Shipley Collins Award for meritorious service to phycology and to NEAS was presented at the banquet to Bill Johansen, who received a Swiss army knife, a head-mounted light and a certificate (Table 12). The DC made available for purchase the first publication sponsored by NEAS: "*The NEAS keys to Benthic Marine Algae of the Northeastern Coast of North America from Long Island Sound to the Strait of Belle Isle,*" a 161 page document edited by Jim Sears and published in 1998. Proceeds from the sale of this book and other phycological memorabilia at a post-banquet auction provided support for the DC student fund. Thirteen students received funds from the DC to help them to attend this symposium. Several companies and organizations presented information on their products or services: Connecticut Valley Biological Supply, Inc., Swift Instruments, and the Rhode Island Natural History Survey. Some of these firms provided door-prizes to students whose names were randomly drawn. Soon after the symposium ended all attendees were sent an email survey asking them to compare the Woods Hole and Plymouth venues. A large majority of the responders favored Plymouth and a decision was made to return to the Sheraton Inn in Plymouth the following year.

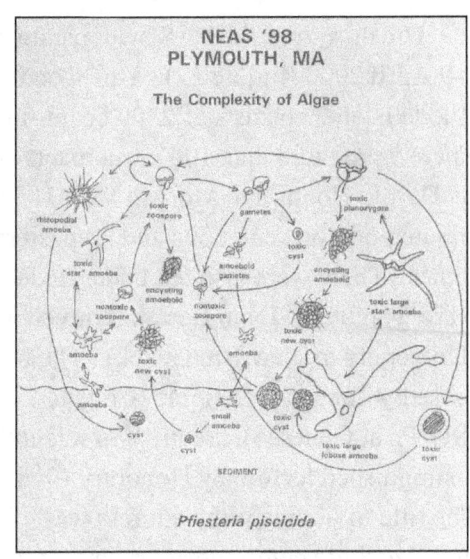

Figure 20. Logo for the 37th symposium (1998). The complexity of unicellular algae as shown by the forms of *Pfiesteria piscicida.* Provided by JoAnn Burkholder.

The thirty-eighth symposium, convened by Peg Van Patten and Gary Saunders, was held at the Sheraton Inn in Plymouth, MA, 16-18 April 1999, and drew 110 registrants. The distinguished speaker, Curt Pueschel (SUNY, Binghamton), gave a fascinating lecture on "*Beyond light: 50 years of contributions of electron microscopy to phycology.*" A mini-symposium featured Harold Hoops (SUNY, Geneseo), on *"(Ultra) structure-function relationships in the green algal flagellar apparatus,*" Pete Siver (Connecticut College), on "*The tale of a scale: morphological and ecological aspects of synurophytes as told by electron microscopy,*" and Richard E. Triemer (Rutgers University), on "*Advantages of freeze substitution in examining euglenoid ultrastructure.*" The Robert T. Wilce awards went to Robert Verb (Ohio University) and Sean P. Grace (University of Rhode Island). For the first time a President's Award was given specifically for the best undergraduate presentation; it went to Amy E. Cocina (SUNY, Geneseo). The Frank Shipley Collins Award went to Bea Robinson (Le Moyne College), who served exceptionally well as NEAS secretary/treasurer from 1992-1997, and was a joy to meet at the symposia through the years. The symposium included a Friday evening collecting trip led by Jim Sears that provided specimens for display. An "Editor's corner," a discussion about the ins and outs of publishing phycological papers, was led by Curt Pueschel (at the time editor of *Phycologia*) and Susan Brawley (at the time editor of the *Journal of Phycology*). The symposium logo was a drawing of the siliceous flagellate *Mallomonas* contributed by Pete Siver and drawn by John Glew (Figure 21).

Figure 21. Logo for the 38th symposium (1999). A drawing of *Mallomonas*, a scaled chrysophyte. Contributed by Peter Siver, drawn by John Glew.

The thirty-ninth NEAS symposium was held at the Whispering Pines Conference Center in Rhode Island on 7-9 April 2000 (Figure 22). This woodland venue, also known as the W. Alton Jones campus of the University of Rhode Island, consists of 2300 beautiful acres of woods, fields, and lakes. There were more than 100 participants at the symposium. It was convened by Harold Hoops and Morgan Vis. The scientific program included 29 contributed oral presentations and 23 posters on a wide range of phycological topics. There were also ten undergraduate contributors. The Wilce awards were won by M. Lynn Berndt (University of Maine) and Jennifer Dalen (University of New Brunswick). The best undergraduate presentations were those by Tomas A. Bonome (Colgate University) and Andrea Shelley (SUNY at Geneseo). Saturday's scientific sessions were concluded with the distinguished lecture by Honorary Chair Sarah P. Gibbs (McGill University). The title of her inspiring address was "*The evolution of peridinin-containing dinoflagellate chloroplasts: surprises and answers.*" Gibbs was also presented with the Phycological Society of America Award of Excellence. A minisymposium on freshwater ecology featured invitees Bob Sheath on "*Tundra stream macroalgae of North America,*" Kyle D. Hoagland, on "*Pesticide impacts on freshwater algal communities,*" and John Wehr on "*How do we study rare freshwater algae?*" The Frank Shipley Collins

Figure 22. Logo for the 39th symposium (2000). Wharf piling and three species of unicells. Drawn by James (Todd) Harper.

award for meritorious service to NEAS and to Phycology, was given very appropriately this year to Jim Sears. The logo, designed by Todd Harper, appeared on the cover of the abstracts booklet, the T-shirts, and the coffee mugs (Figure 22). As has recently become customary, a post-banquet auction (auctioneer Glen Thursby) provided funds for the DC student fund that help students attend the annual symposia.

Figure 23. Logo for the 40th symposium (2001). A montage of several algae. Artist unknown.

The fortieth NEAS symposium was again held at the Sheraton Inn in Plymouth, Massachusetts, on 20-22 April 2001 (Figure 23). There were more than 100 participants, including some members of the PSA Executive Committee. The symposium was ably directed by co-conveners Karen Culver-Rymsza and Carl Grobe. The Wilce award winners were Penny Humby (University of New Brunswick) and Xingye Yang (SUNY at Syracuse). Two undergraduate students shared the President's award: Julie Price (University of New Brunswick) and Jennifer Sallee (Roger Williams University). Saturday's scientific sessions were concluded with the distinguished lecture by Ian Davison (University of Maine) who presented a very interesting talk on "*Travels with PAM: Insights into the strange world of seaweed physiology.*" There was also a special session on *Algae and Technology* as part of the meeting, as well as a collecting trip. The Frank Shipley Collins Award for meritorious service to NEAS and to Phycology went to Bob Wilce. As we are all aware, Wilce had been a driving force in most all aspects of NEAS.

The forty-first NEAS symposium was held 20-21 April 2002 at the University of New Hampshire in Durham (Figure 24). Organizers of this year's

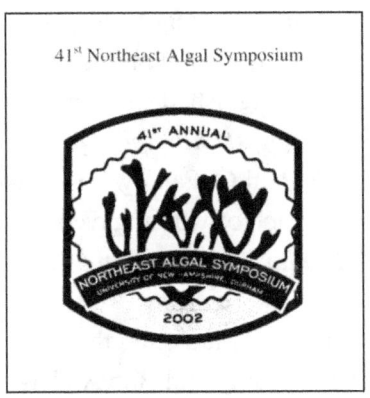

Figure 24. Logo for the 41st symposium (2002). Using *Fucus* as the model, Tuyen Nguyen Teasdale's design was originally used for "Seacoast Seaweed Ale," a homebrew project with husband Brian Teasdale (University of New Hampshire).

symposium were Anita Klein, Jack Holt, and Chris Neefus. Almost 100 attendees were present. The scientific program consisted of 20 platform presentations and a similar number of posters. Colin R. Bates (University of New Brunswick) and Aaron Wallace (University of New Hampshire) won the Wilce awards for their presentations. The President's Award was won by Anthony Gallo (SUNY Geneseo). The Frank Shipley Collins award for service to NEAS and phycology went to Frank Trainor who coincidentally was honored with the PSA Award for Excellence. Bill Johansen listed Frank's contributions to NEAS and some of his former students told about their friendship with him and how much they appreciated his mentorship: Pete Siver, Ed Boger, Elliot Shubert, Ed Bonneau, and Eduardo Morales. Saturday's presentations were concluded with the distinguished lecture by Bob Wilce with the intriguing title "*Musk oxen, fine structure and molecular biology.*"

The forty-second NEAS symposium was held on the rainy weekend of 25-27 April 2003 at Skidmore College, in historic Saratoga Springs, NY. David Domozych and Gary Saunders organized the symposium, and all attendees proclaimed it to be a very successful event. The scientific program consisted of 22 oral presentations and 25 posters. Saturday's presentations were concluded with an extremely interesting distinguished lecture by Ralph Quatrano (Washington University), entitled "*Cytoskeletal and cell wall interactions during the establishment of cell polarity in the* Fucus *zygote.*" Highlighting the banquet presentations were the Robert T. Wilce and Presidential awards given to students with the best talks and posters. In the oral presentation category, Brian Teasdale (University of New Hampshire) received the Wilce award. In the poster category, the Wilce award went to Susan Clayden (University of New Brunswick). The President's Award, given for the best undergraduate presentation, was shared by Hannah Shayler (Connecticut College) and Martin Monahan (Eastern Connecticut College). The Frank Shipley Collins award for service to NEAS and phycology was presented to the well-known Craig Schneider. Bill Johansen, Bob Wilce, and Jim Sears listed Craig's many contributions to NEAS. Schneider also served as auctioneer following the award presentations. The Sunday morning sessions included an interesting mini-symposium, "*Microscopy Imaging and Phycology*", featuring Mark Farmer (University of Georgia), Michael Gretz (Michigan Technological University), and Stephen Droop (Royal Academy-Edinburgh). Invited lectures dealing with the biology of the nearby Hudson River were presented by Jonathon J. Cole (Cary Arboretum, Millbrook, NY), and on Antarctic freshwater algae by Ray Kepner, Jr. (Marist College). Sunday morning sessions also included laboratory workshops on microscope techniques.

The forty-third symposium was held 23-25 April 2004 at the scenic Avery Point Campus of the University of Connecticut. The hosts, Louise Lewis, Senjie Lin and Charlie Yarish, were roundly praised for organizing an excellent symposium that attracted 132 registrants, 69 of whom submitted abstracts. Long-time University of New Hampshire phycologist Art Mathieson was the honorary chair. Notable contributions were made by several sponsors: Project Oceanology, an organization that provided free dormitory lodging for all student attendees; the Connecticut Sea Grant College Program's award that allowed NEAS to bring in four speakers for a Sunday morning mini-symposium; and the Swift Microscope Company that raffled off a field microscope. As usual, at the heart of the symposium were oral presentations and posters prepared by students and professional scientists alike. Andrew Bamburger (Great Lakes Institute of Ecology) and Karen Pelletreau (University of Delaware) won the Wilce awards for their presentations. The President's Award was shared by David Sakoda (Wheaton College) and Megan Brennan (Susquehanna University). The theme of the 2004 minisymposium, "*Algae and human affairs: making connections from genes to ecosystems*," was admirably addressed by this year's distinguished speaker, Patricia Tester (NOAA, National Ocean Service) in her lecture "*Copepodology for the phycologist with apologies to G.E. Hutchinson*." Sunday's minisymposium featured Greg L. Boyer (SUNY, Syracuse), Paul G. Falkowski (Rutgers University), James T. Carlton (Williams-Mystic) and Erick Ask (FMS Corporation), each of whom gave a 30 minute lecture. The lunch and business meeting was followed by a workshop on "*Molecular techniques in phytoplankton research*."

The forty-fourth annual NEAS Symposium was convened at Samoset Resort in Rockland, Maine, on the weekend of 15-17 April 2005. The symposium, under the capable leadership of Charles O'Kelly and Bob Anderson, began with tours of the FMC Marine Colloids facilities. During the weekend the scientific program consisted of 29 oral talks and the remarkably large number of 51 posters. Bob Vadas (University of Maine) was the honorary chair of the symposium. Two Wilce awards were given for the best oral presentations: Priya Sampath-Wiley (University of New Hampshire) and Jeremiah Hackett (University of Iowa). In the poster category Lisa Pickell (University of Maine) won the Wilce award. The President's Award for the best presentation by an undergraduate student went to Katie Williams (University of New Hampshire). The distinguished speaker was Debashish Bhattacharya (University of Iowa) with a lecture entitled "*A phylogenetic tree and genomic perspective on the origins of photosynthetic eukaryotes within the tree of life*." Also, this weekend the Frank Shipley Collins award for outstanding contributions to NEAS was presented to the effervescent former NEAS chairperson, Larry Liddle (Long Island University).

Logos from 2003, 2004, and 2005 are shown in Figure 25.

In 2006, the forty-fifth NEAS Symposium was held in 21-23 April at Marist College on the shores of the Hudson River in Poughkeepsie, New York. The three co-conveners were Ray Kepner, John Heimke and David Domozych. The theme of the symposium was '*Algal Biofilms*.' Twenty-one oral platform presentations and 32 posters constituted the heart of the scientific program. Bill Johansen was the honorary chair of the symposium. The Wilce Award for the best oral presentation by a graduate student went to Hilary A. McManus (University of Connecticut). Daniel McDevit (University of New Brunswick) received the Wilce

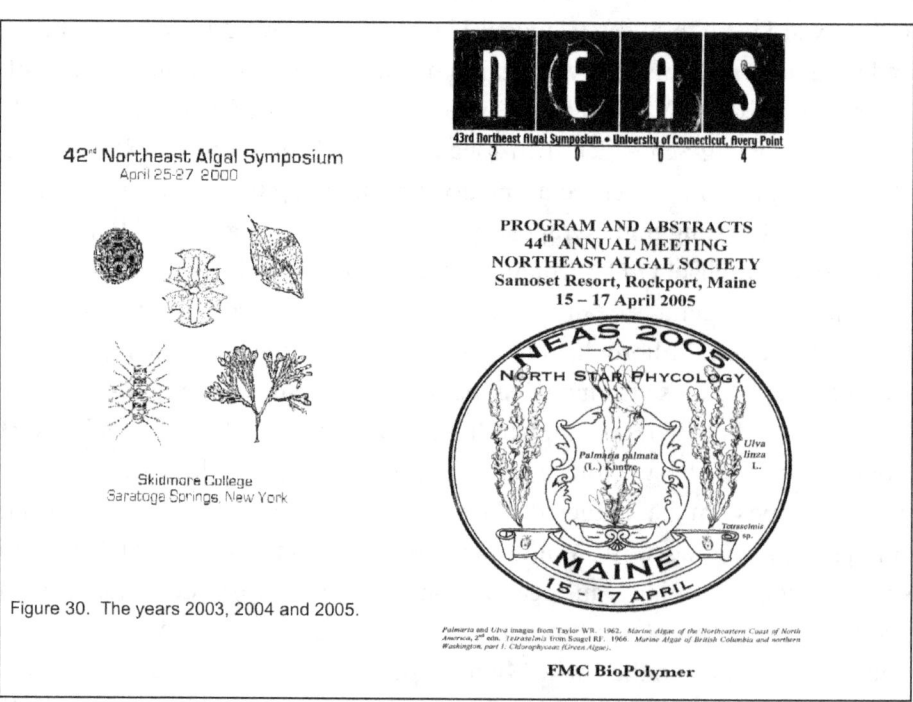

Figure 25. The logos from years 2003, 2004, and 2005.

award for the best poster by a graduate student. The President's Award for the best presentation by an undergraduate student went to Cayelan Carey (Dartmouth College). Graham Underwood (University of Essex) was the distinguished speaker and gave an excellent lecture entitled *"Life in estuarine biofilms: small-scale complexity and large scale importance."* Curt Pueschel was awarded the Frank Shipley Collins award for his many contributions to NEAS over the years.

The forty-sixth NEAS Symposium in 2007 was held at the Village Inn at Narragansett Pier Hotel and Conference Center in Narragansett, RI. The organizers, Morgan Vis and Glen Thursby, convened the symposium during the weekend of 20-22 April 2007. The theme of the symposium was stated as a question, *'Where do we go from here?'* As usual, the graduate students chosen for the best oral and poster presentations were presented with Robert T. Wilce Awards. In the oral presentation category, Jessica Muhlin (Maine Maritime Academy) won as did Malcolm McFarland (University of Rhode Island) in the poster category. As part of the award Jessica and Malcolm were each given a $500 travel grant to attend the upcoming PSA meeting, and a $75 check. The President's Award, for the best undergraduate presentation, went to Laura Lambiase (Skidmore College). She received a check for $50. The distinguished lecture was given by Dennis Hanisak (Harbor Branch Oceanographic Institution); it was entitled *"Heroes in the seaweed: John H. Ryther."* As usual, Glen Thursby ran an auction after the banquet. Glen, a lively auctioneer, gave students a 50 percent handicap at the auction, that is, they paid only half of their bid if they win an item. He continued this practice as long as he remained auctioneer.

In 2008, co-conveners Chris Neefus and Anita Klein organized the forty-seventh NEAS Symposium, and NEAS returned to the University of New Hampshire for 18-20 April 2008. The theme for the weekend was *'The Globalization of Phycology via the Internet.'* The Robert T. Wilce Awards went to Yen-Chun Liu (Northeastern University) and Nathan J. Smucker (Ohio University) for their oral presentations. Jennifer Day (University of New Hampshire) won the Wilce Award for her poster. Winners of Wilce awards serve one-year terms on the EC, and receive up to $500 each for travel to the upcoming PSA meeting. The keynote speaker was Michael Guiry (Martin Ryan Institute, NUI, Galway, Ireland) whose interesting talk was entitled *"AlgaeBase."* In AlgaeBase, Mike and his collaborators have established an electronic data base providing a taxonomic backbone of algal taxa that can be tapped into by researchers world-wide. Auctioneer Glen Thursby reported that the post-banquet auction yielded about $1750 for the DC. This year a new session called *'Phyco-Speed Dating'* was introduced and moderated by Jessica F. Muhlin (Maine Maritime Academy). It was a fast-paced series of presentations aimed at stimulating discussion, questions, and feedback on ideas for pet projects that participants would present. The 2-3 minute presentations were followed by 2-3 minutes of questions and comments. The membership director, Brian Wysor, worked to trimming the electronic mailing list to eliminate people who had moved on.

The logos for 2006, 2007, and 2008 are shown in Figure 26.

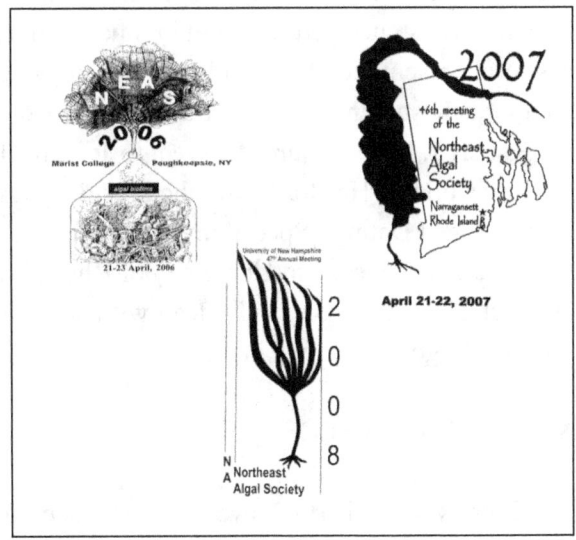

Figure 26. The logos from years 2006, 2007, and 2008.

The forty-eighth NEAS symposium was convened 17-19 April 2009 at the University of Massachusetts in Amherst. The conveners were three long-time members of NEAS: Bob Wilce, Craig Schneider and Bill Johansen. The honorary chair for the symposium was Frank Trainor. The theme of this symposium was '*Algae and Biofuels*' and the keynote speaker was Scott W. Gordon, CEO of Green Technologies, LLC, Winooski, Vermont. The title of Scott's interesting lecture was "*Small scale biodiesel production (including some dos and don'ts for making biodiesel from algae)*." There were about 120 attendees. The Robert T. Wilce awards went to Susan L. Clayden and Bridgette Clarkson, both from the University of New Brunswick. Elizabeth Sargent (Roger Williams University) won the President's Award for the best presentation by an undergraduate student. A session on Phyco-Speed Dating was coordinated by Hilary A. McManus. In addition, two renowned phycologists were invited to lecture at the symposium: Gary W. Saunders (University of New Brunswick) on "*The promises and pitfalls of molecular-assisted alpha taxonomy*" and Mike Wynne (University of Michigan) on "*Marine algae and early explorations in the upper North Pacific and Bering Sea*."

On 16-18 April 2010, the forty-ninth NEAS symposium, the last one before going back to MBL at Woods Hole, was convened by Brian Wysor and Chris Lane at Roger Williams University in Bristol, Rhode Island. The theme was '*Algal biodiversity: shifts in algal distribution*.' There were 110 registrants. Dr. Olivier De-Clerck, from Ghent University, Belgium, was the keynote speaker with a talk entitled "*Species, patterns and algae: a phylogenetic perspective*." However, DeClerk's lecture was delivered from his Belgian homeland via the internet. His trip to the NEAS symposium was cancelled because of a massive volcanic eruption in Ice-

Figure 27. Logos for the 2009 and 2010 meetings.

land whose ash negated jet travel in much of northern Europe and the North Atlantic. The Wilce Award for the best oral presentation by a graduate student went to Molly Letsch (University of Connecticut). For the first time, the graduate student preparing the best poster was awarded the Frank R. Trainor Award (replacing in name only the Robert T. Wilce Award). It went to Jeremy Nettleton (University of New Hampshire). Elisabeth Cianciola (Trinity College, Hartford) won the President's Award. Hilary A. McManus again coordinated a session on Phyco-Speed Dating. Special invited lectures were also presented by Marcie Marston (Roger Williams University), Tatiana Rynearson (University of Rhode Island), Craig Schneider (Trinity College) and Peter Siver (Connecticut College). These talks dealt with population genetics and the biogeography of viruses and various algal groups. Logos are shown in Figure 27.

NEAS COMMITTEES

Every year until 1982, two scientists (occasionally three) took it upon themselves to convene a NEAS symposium for the following year. This worked in a hit or miss fashion fairly well, but there was always the nagging possibility that one year it would fail when, for one reason or another, the conveners could not adequately plan a symposium. The preparation of a symposium suddenly, and perhaps unexpectedly, imposed a large work-load on two scientists already buried in teaching and research. Moreover, several thousand dollars in bits and pieces had to be kept track of. Meticulous records had to be kept and plans had to be carried out. Meanwhile the NEAS symposia were becoming ever better attended. Something could go awfully wrong. Something had to be done.

   An Ad Hoc Committee. Flurries of letters and phone calls in the summer and early fall of 1981 resulted in a seminal meeting of 16 phycologists from New England and eastern Canada at the Marine Biological Laboratory in Woods Hole on 21 November 1981 to discuss the future of NEAS (Table 7). This meeting was called by Bob Wilce in response to the increasing feeling that NEAS could not continue to grow and maintain a string of successful annual symposia unless some changes were made. The consensus of the group was that NEAS was a healthy entity that should be shored up. Its strengths outweighed its weaknesses. It was a positive force in promulgating phycology and providing mechanisms that facilitated interactions among phycologists, both students and professionals.

Table 7. *Members of the sixteen-person ad hoc committee that set the groundwork for NEAS. The list was generated from the memories of several of the members. The name of the sixteenth member will be added when the memories permit.*

Bud Brinkhuis
Don Cheney
Annette Coleman
Marilyn Harlin
Peter Heywood
Llewellya Hillis-Colinvaux
Bill Johansen
Ray Jones
Norm Lazaroff
Tom Lee
Joe Ramus
Jim Sears
Frank Trainor
John van der Meer
Bob Wilce

   The ad hoc committee prepared a report dated 13 April 1982 that included its recommendations; it was authored by Jim Sears, Marilyn Harlin and Bob Wilce. In it, the committee clearly saw the need for organized continuity from year to year, something that had largely been lacking. The report stated that change was needed: "...*to maintain the status quo will not assure the continuance of the positive aspects of the annual meeting.*" The recommendations were to be put to the membership at the annual business meeting held during the 1982 symposium where the members would provide input to the committee. In addition to the problem of continuity, the ad hoc committee also recommended that NEAS become affiliated with the national phycological organization, the Phycological Society of America (PSA). This issue is considered on the next page.

The ad hoc committee recommended the establishment of an Executive Committee (EC) consisting of five officers and three members-at-large. The officers were to be two chairpersons, two vice-chairpersons and one secretary/treasurer. The chairpersons would run the annual symposium (now called co-conveners) and the vice-chairpersons would serve one year after which they would become chairpersons for one year. The secretary/treasurers would serve five-year terms to insure continuity in financial matters. The members-at-large would serve staggered terms of three years each, again to provide continuity and stability.

During its meeting in November 1981, the ad hoc committee recommended the establishment of a five-person nominating committee for one-year terms to solicit nominations and carry on an election at the symposium. (This was later reduced to three members.) In order to be prepared for the hoped for acceptance of the plan by the membership at large (in May, 1982), the ad hoc committee selected the first nominations committee and named Dennis Hanisak as its chair. Hanisak's subsequent report made some useful suggestions about how this committee should be organized. This nominating committee prepared a slate of candidates for the business meeting to be held at the annual symposium in May, 1982. The slate of candidates was as follows.

Chairpersons: Bud Brinkhuis, Peter Heywood
Vice-chairpersons: Don Cheney, Llewellya Hillis-Colinvaux
Secretary/treasurer: Bill Johansen
Members-at-large: Marilyn Harlin, Frank Trainor, Bob Wilce
Nominations Committee (to prepare a slate of candidates for the 1983 symposium): Tom Lee, Bea Robinson, John van der Meer (chairperson)

At the business meeting additional nominations were requested from the floor and, since there were none, the complete slate of candidates was elected by acclamation. Thus, in May, 1982, the first NEAS Executive Committee (EC) was established.

A functioning Executive Committee. In the years since 1982, the EC has performed exceedingly well. However, with time there have been modifications: (1) The committee is now led by a President (Table 8), (2) The position of Secretary/Treasurer has been separated into two positions, (3) A Development Committee (DC) now augments the Executive Committee, (4) the co-conveners each serve for three years, one before they organize the meeting and one after, (5) there are now two members-at-large instead of three, (6) a new post, Vice-president/President-elect, has been established, (7) graduate student award-winners from the previous symposium now serve one-year, non-voting terms on the EC.

The first meeting of the NEAS Executive Committee, on 6 July 1982, achieved notable results. (1) The EC voted to have a single Chairperson, now called the President, who would serve for a two-year term; Bob Wilce was elected by the EC to serve in this position until the NEAS membership could vote for officers at the 1983 symposium. (2) The pros and cons of becoming affiliated with PSA were discussed with the help of information provided by Annette Coleman, President of PSA (considered in detailed later). (3) A draft of a NEAS bylaws document, prepared by Bob Wilce, was examined and revised; after revision it was still considered a draft document (see Appendix B).

Since 1982 the EC has met annually, usually in the fall, to discuss and plan a myriad of issues, many related to the upcoming symposium in the following spring. The EC also meets during the annual NEAS symposium.

The President also presides over a business meeting of the entire NEAS membership, usually on the Sunday of the symposium weekend.

Other members of the EC are two members-at-large. Like the other officers, they are also elected when needed by the membership at the annual symposium business meeting. The members-at-large serve staggered three-year terms. They serve as sounding boards for ideas expressed at the annual fall EC meetings.

In 1986, graduate students were added as one-year, non-voting members of the EC. These would be the students who received the Wilce or Trainor Awards at the previous symposium.

The formal NEAS bylaws document written by Bob Wilce in 1982 was accepted in 1983. In subsequent years, it has been modified several times (see Appendix B for the latest version). In legalese, the NEAS Bylaws consist of ten articles entitled Name, Purposes, Membership, Meetings, Executive Committee, Committees and Student Representation, Election of Officers, Amendments, Nonprofit Status and Dissolution.

NEAS officers. During the 1990s, the duties of NEAS officers were spelled out (Appendix C). Since the EC was established 29 years ago, ten phycologists have served as President of the NEAS (Table 8). In 1982, Bob Wilce was selected as the NEAS Chairperson for the first year, and at the symposium in 1983 he was elected for a two-year term by the membership. Except for Bill Johansen, Craig Schneider, Curt Pueschel and John Wehr, the other presidents served for two-years each. Johansen served for five years, four of them based on his election for two year terms in 1987 and 1990. Between these terms he presided for one year (1989) when President-elect Bud Brinkhuis passed away. Craig Schneider and Curt Pueschel were also elected for two-year terms twice. John Wehr served a three-year stint.

Perhaps the most significant duty of the President is to preside at the Fall meeting of the EC, an important planning session. He/she also presides at the business meeting held annually at the symposium. Other duties of the president are outlined in Appendix C. The position of Vice-president/President-elect was established in 2002. This individual serves one year as Vice-president and then two years as President.

Table 8. *The Chairpersons (Presidents) of NEAS since 1982 when elections were first held. See Appendix D for full names and affiliations.*

1982-1985	Bob Wilce
1985-1987	Annette Coleman
1987-1992	Bill Johansen
1992-1994	Larry Liddle
1994-1996	Susan Brawley
1996-2000	Craig Schneider
2000-2004	Curt Pueschel
2004-2006	Gary Saunders
2006-2009	John Wehr
2009-2011	Morgan Vis

(Bud Brinkhuis was president-elect for 1989-1991 but he died unexpectedly before he could serve his two-year term.)

The position of Secretary/Treasurer was designated as an office with a term of five-years, and three persons have served in that capacity: Bill Johansen (1982-1987), Craig Schneider (1987-1992) and Bea Robinson (1992-1997). Unfortunately, Bea became ill in 1995 and Peter Bradley (who had been elected as co-chair of the nominations committee with Lucie Maranda) was asked by the President, Susan Brawley, to be Interim Secretary and to assist with the preparation of an officers manual (Appendix C). In 1996, the EC recognized the large amount of work involved in being secretary/treasurer and decided to split the position. The membership then elected Peter Bradley for a five-year term as Secretary (1996-2001) while Bea Robinson finished her term solely as Treasurer one year later. In 1997, Glen Thursby was elected Treasurer for five years, serving until 2002. Greg Boyer then was elected Treasurer twice, serving five-year terms from 2002 to 2012.

After serving one five-year term as Secretary, Bradley was re-elected for a second five-year term at the symposium in 2001. He thus served as secretary for ten years before being replaced by Louise Lewis in 2006. Her term expires in 2011. The duties of the Treasurer and Secretary are outlined in Appendix C.

Although the roles of the co-conveners have changed slightly through the years, they are essentially the same as they were in the 1990s. The co-conveners have the direct responsibility for the upcoming symposium, a big job with many aspects. Fortunately, they know they will have the help of other members of the EC when needed. Moreover, the co-conveners for an upcoming symposium have already been members of the EC for the prior year, and thus they know ahead of time much that they need to plan for. The extensive duties of co-conveners are listed in Appendix C.

Nominations Committee. At the heart of an EC representing the entire membership of NEAS is a fair and simple system for nominating and electing officers. After the first Nominations Committee was selected by the ad hoc committee of 1981-1982, and the committee size was reduced from five to three, a new committee has been constituted every year. Until recently the elected chairperson of each new Nominations Committee selected two other persons to serve on the committee during the months that follow. Prior to each symposium a call for nominations is mailed out with the first mailing. Nominations are solicited for offices that will become vacant at the upcoming symposium, and for a new Nominations Committee chairperson. After receiving the nominations, the three-person committee prepares a slate of candidates, including space for additional nominations. The elections were sometimes by mail ballot (sent with the second mailing) before the symposium convenes, or by paper ballot just before the symposium business meeting.

At a vote taken at the all-symposium business meeting in 2002 the makeup of the Nominations Committee was changed. In the future, the membership would elect a chair and a chair-elect, each to serve two-year terms. The chair will choose a third member who will serve for one year. Moreover, the NEAS President shall serve as a non-voting member of the Nominations Committee.

Publications Committee. A "*Flora Committee*" comprised of Jim Sears, Barry Colt and Bob Wilce was established by the EC in 1987, and subsequentially it was renamed the "*Publications Committee*." This group was asked to explore the idea of preparing regional floras for both freshwater and marine algae of the western North Atlantic region. These projects could include monographs, occasional papers and a comprehensive database of phycological literature that had been assembled by Barry Colt.

In 1990, the Publications Committee suggested that NEAS should develop identification keys for the benthic seaweeds of the Northeast Atlantic. Accordingly, Jim Sears began compiling work on species from Newfoundland to Long Island Sound. Eventually, four New England Sea Grant programs offered financial support: the University of Connecticut, Massachusetts Institute of Technology, Woods Hole Oceanographic Institute and the University of Maine/University of New Hampshire Sea Grants. In addition, Southeastern Massachusetts University also made a financial award to Sears, who was the project's principal author and editor.

Lists of taxa to be included in the key were sent to various phycologists for their comments and an early draft was placed on display at the thirtieth symposium (1991). Several phycologists agreed to work on parts of the draft: Charlie Yarish (small greens, Ctenocladales), Art Mathieson (geographic distribution), Carolyn Bird (selected northern taxa), Peg Van Patten (editing), Milan Keser (distribution and phenology), Bob Steneck (Corallinales), Bob Wilce (Phaeophyta), Martine Villalard-Bohnsack (Ulvales), Craig Schneider (Rhodophyta, Xanthophyceae) and Bob Vadas (Laminariales).

Reports that Sears prepared for the EC clearly showed the tremendous amount of work and planning that went into this project. In 1994, a draft of a literature guide to the Gulf of Maine was prepared in cooperation with scientists at the Bigelow Laboratory for Ocean Sciences. A comprehensive database prepared by Barry Colt on New England algae covering the years 1829-1984 included pertinent literature and collection sites.

In 1995, the Publications Committee suggested that publications might include floristic analyses of particular regions, historical perspectives of algal studies, and economic analyses. The idea was that such publications would serve to encourage an awareness and understanding of the importance of algae.

By 1995 Sears reported that portions of the keys were ready for review. The '*NEAS keys to the Benthic Marine Algae of the Northeastern Coast of North America from Long Island Sound to the Strait of Belle Isle*' was published by NEAS in 1998 (Sears 1998) as NEAS Contribution Number 1. The volume, 161 pages long, included a series of keys to aid in identifying all the known species of seaweeds between Long Island and Newfoundland. In addition to the aforementioned phycologists, the following contributed to the volume: Carolyn Bird, David Garbary, Joan Hare, Edward Hehre, Christiaan van den Hoek, Robert Hooper, Art Mathieson, Ruth Nielson, Poul Pedersen, Craig Schneider, Bob Steneck, Bob Vadas, Martine Villalard-Bohnsack and Bob Wilce.

The frontispiece of the book was a colorful watercolor by Mary Elizabeth Gordon entitled '*The Shallow Sublittoral*.' Fifty enlarged copies of this painting were printed as a poster. The book and poster were placed on sale at the thirty-seventh symposium in 1998. Five hundred copies of this edition were sold. Profits from the sales went into the DC fund. Christine Maggs published a generally positive review of this document in the *Journal of Phycology* (Maggs 1998).

Four years later, Sears updated and corrected the NEAS keys (Sears 2002) as NEAS Contribution Number 2. New in this second edition were five excellent plates of color illustrations of common seaweeds taken by Jim Sears. The final book was a beautiful, professional publication with a comb binding.

The work by Jim Sears has been used as a reference in the '*Digital Herbarium*' established by the MBL-WHOI Library. This digital herbarium is a fully searchable, online database of the marine algae in their library. It provides specimen images and associated data.

In the early 2000s, Bill Johansen and Peter Bradley (then NEAS secretary) began preparing a paper on the history of NEAS. In 2003, this ended up as NEAS Contribution Number 3, a document on the history of NEAS through 2002 (Johansen & Bradley 2003). The present document incorporates the many developments that have occurred since then, ending with the forty-ninth symposium in 2010.

Affiliation with the Phycological Society of America. One of the recommendations made by the ad hoc committee of 1981-1982 was for NEAS to investigate the feasibility of becoming affiliated with the Phycological Society of America (PSA). Annette Coleman, then president of PSA, facilitated this action by drafting a proposed amendment to the PSA bylaws that would be considered by the Executive Committee of that body in August, 1982. If approved, the PSA membership would vote on the amendment by mail ballot. This amendment allowed for the inclusion in PSA of affiliate societies, such as NEAS. There would be no fiscal obligation in either direction. The advantages to NEAS in formally becoming "The Northeast Section of the Phycological Society of America" would be a more recognized status that might be advantageous when seeking funding for symposia and the like. We would still informally be called NEAS. The completion of this action has had minimal implications for NEAS, but it may become important in the future. The NEAS EC stressed in 1982 that, in spite of an affiliation with a national organization, NEAS will strive to maintain the informality and friendliness that have hitherto characterized our symposia. In recent years members of the Executive Committee of the PSA have attended NEAS symposia. The PSA has been very supportive of NEAS and recently has provided sets of the *Journal of Phycology* for the DC to sell, the proceeds to assist students interested in phycology.

THE SYMPOSIA

The logistics of organizing the symposia are large. Based on the more recent symposia, a successful pattern evolved. Many attendees arrive at the symposium venue Friday evening to register and to socialize at an evening mixer. After a breakfast on Saturday morning and registration for attendees as they arrive, one of the conveners opens the meeting with a few remarks. By about 9 AM, fifteen minute oral presentations by students begin. After a mid-morning coffee break more student oral presentations occur until noon. Lunch for attendees and a meeting of the EC take up an hour and a half or so at mid-day. After more student talks and another coffee break in mid-afternoon, a half-hour session on Phyco-Speed Dating is convened. In the late afternoon, the distinguished speaker gives his/her talk. Typically, a session for viewing posters follows the talk. The final events on Saturday include an hour-long social gathering and a silent auction, and then the banquet at 7 PM. After the feasting, various awards are presented and the evening ends on a gala note with the live auction.

Following a continental breakfast on Sunday morning, the second day's contributed papers are given, now mostly by professional scientists, followed by a mid-morning coffee break. A special program sometimes follows the break. Some years, such as in 2010, the special program consisted of four invited papers, each a half-hour long. The symposium concludes shortly after noon with the annual NEAS business meeting and box lunch.

For many years the EC has held an important planning meeting during the fall of the year. This is a time when ideas are evaluated and plans are made for the upcoming April symposium. Following the fall EC meeting the co-conveners take the ball and run with it. They deal with the symposium site officials and, at the appropriate time, send out circulars via email to those on the membership list. Typically one of the converers deals with the scientific program, organizes abstracts and prepares an abstract booklet. The other convener collects and counts the incoming payments, and deals with the details of housing at the venue. Also, during these pre-symposium months the DC arranges the many details that will ensure an income to the student fund. Usually, this consists of having mugs and T-shirts prepared and emblazoned with the current logo. Also, items to be auctioned off after the banquet were solicited with the meeting announcements early in the calendar year.

Symposia locations. The Woods Hole years, 1976-1997, were good ones for NEAS. The MBL venue year in and year out during the Society's early years acted like a cement that gave the symposia stability. The symposia increased in size and scope during this time, and some members could not imagine that we would meet elsewhere. After a long and fairly comfortable relationship with the MBL and the Swope Conference Center, that relationship became increasingly more difficult for NEAS, with imposed new charges and restrictions. In 1997, the EC had a difficult decision to make: either remain at Woods Hole within the new restrictions or return to the earlier model of moving about to different venues. The latter prevailed.

Some of the problems with returning to Woods Hole were that (1) MBL was loathe to provide accommodations on Friday night before the symposium officially opened; (2) they required a deposit of $1000 to be paid months before the symposium convened; (3) they refused to provide lunch on Sunday; and (4) there was no MBL representative on site during the symposium to provide assistance when needed. These problems made it vexing for co-conveners to organize and smoothly run symposia. In the summer of 1997, President Craig Schneider wrote letters to the director of the Swope Conference Center, the MBL Executive Director, and the chairman of the Executive Committee of MBL. This was an unsettling time for the EC as no reply was received until the following year. Therefore, in 1998 the first symposium since 1975 was held in a location different from Woods Hole, namely at the Sheraton Inn in Plymouth, Mass. Although sentiment favoring the Woods Hole site remained strong among a selected few attendees, in 1998 most voiced satisfaction with the Plymouth site. Although the idea of returning to Woods Hole has cropped up time and time again, until now (2011) all subsequent symposia have been convened elsewhere (Appendix A).

Mailings. For many of the early years a mailing list of attendees at previous symposia was maintained by one of the officers. For several years Bud Brinkhuis and later, Bill Johansen, kept and updated the list and regularly supplied mailing labels to the appropriate co-convener at the appropriate time. Each year the membership was informed of the upcoming symposium by two mailings (circulars), the first in December and the second in February. The first mailing typically contained an announcement and news about the upcoming symposium and a call for nominations. The second mailing contained (1) more news and announcements, (2) a registration form, (3) a message from the DC soliciting donations from senior phycologists, (4) a call for proposals for financial aid from student attendees, and (5) a call for abstracts. Sometimes, a third limited mailing notified presenters of the schedule.

In 1995, the EC selected Bill Johansen to be NEAS Membership Director and Archivist, a new post (Appendix C). The principal task in this position was to maintain the mailing list and distribute mailing labels at appropriate times to the co-conveners about to send out mailings. Johansen also kept records of NEAS business, and these records have proved to be valuable in preparing the present document. Due to unforeseen, circumstances Johansen resigned as membership director in 2000. The mailing list was then maintained by Treasurer Glen Thursby until 2002, when Chris Neefus was appointed as membership director for a five-year term. After Neefus, Brian Wysor became membership director, serving from 2007 to 2012.

By 2003, Neefus announced that a large percentage of the names on the mailing list had email addresses. Therefore, in 2004, the announcements went out by email and the names without email addresses went out by snail mail. At present, all announcements are sent by email. The NEAS website, www.e-neas.org was originated in 2008 by, and has been maintained since, by webmaster Nic Blouin (University of Rhode Island).

DEVELOPMENT COMMITTEE

The NEAS Development Committee (DC) was born at an EC meeting on 29 April, 1984. At a post-symposium Sunday afternoon meeting Chairperson Bob Wilce appointed an "Ideas Committee" (IC, later to be renamed the DC) of Craig Schneider, Ed Boger, Pete Siver and Llewellya Hillis-Colinvaux. The goals of this committee were to suggest how money could best be spent (a frequent goal of committees) and to generate new ideas for NEAS. At a later EC meeting in October, 1984, a more precise mandate was established for the IC. It would be a semi-independent committee charged with promoting phycology in the Northeast and supporting NEAS. It would have its own budget—thus, in 1984, the Secretary/treasurer was authorized to provide $1000 as seed money to the IC. It would use various promotional items to build up a substantial fund so that awards of various types could be made to students members (not to faculty or other professionals). To generate money, the DC started by selling T-shirts and mugs with the annual meeting logo, coasters, old reprints and postcards. A few years after its inauguration as a committee, the DC began holding fund-raiser auctions at each symposium.

During the 1985 symposium, the IC sold mugs and T-shirts and received some personal donations. At discussions in the EC meeting of September, 1985, it was decided to change the name to 'Development Committee' (DC) and include a graduate student member. Each DC member would have a renewable, one year tenure. Craig Schneider was chosen to be its first chair. The DC would review continually the development of NEAS activities and bring suggestions to the EC.

The DC has come a long way since its origin in 1984. Three years after its inception (1987) a budget report gave a balance of $2270 in its account ($1000 of which was its seed money). Much of the success of the DC has been due to members that have spent countless hours arranging for mugs, T-shirts, and the like to sell, and other hours soliciting donations and dispensing awards. Some of the most active members of the DC have been Ed Boger, Ed Bonneau, Barry Colt, Craig Schneider, Pete Siver, Glen Thursby, and Brian Wysor.

The mugs, T-shirts, and abstract booklets exhibit the NEAS logo for the symposium year. As far as we can tell the first logo appeared in 1983, and every year since then a different logo has been produced (see Figures). In 1986, the membership voted for a permanent NEAS logo among five artfully produced designs. The winner was a drawing of an apex of *Plumaria elegans,* a New England marine red alga, as seen through a microscope; the artist was Craig Schneider (Figure 28). According to Craig Schneider in 2011 this is presently considered a "temporary placeholder, not a permanent logo."

During the symposia held so far in the twenty-first century an auction has been an exciting event held Saturday evening after the banquet. The auctioneer has usually been the humorous Glen Thursby, who has in various documents been called a "master auctioneer, celebrity auctioneer or an auctioneer extraordinaire." At times, Barry Colt, Tom Lee and Craig Schneider served as auctioneer. A variety of items contributed by symposium attendees have been auctioned off. For example,

Fig. 28. Permanent NEAS logo, an apex of *Plumaria elegans* drawn by Craig Schneider.

reprints, books, long runs of journals, photographs, artwork and algae-related miscellany are the most common items, but non-phycological items have also appeared on the auctioneer's table. Frank Trainor, an expert wood carver, has contributed scores of beautifully carved wooden birds to be auctioned off beginning in 1997, and to date these have generated approximately $6,000 alone for the DC. Peter Bradley's wife, Rosslyn, often brought her exquisite bonsai trees or hand-made scarves to the auction. Typically, in later years, the auction has generatred more than a thousand dollars for the DC at each symposium, and in 2008 it earned the DC about $1800.

In 1994, the DC established its '*Senior Phycologist Award*.' Donations of $100 were solicited from older professional phycologists that could be used to support a student symposium attendee (see Appendix B). All of the DC funds generated by various means is used to help defray more and more student expenses at each annual meeting.

For several years Barry Colt was a vital component of the DC. Unfortunately, Barry passed away in 2006 at the age of 76. Very appropriately, the DC was formally named in Barry's honor, the 'Lebaron Colt III Student Development Committee,' or in its shortened form, the 'Colt Committee.'

FINANCES

The NEAS ad hoc committee of 1981-1982 recommended that one of the NEAS officers should be a Secretary/treasurer. His/her duties would "include maintaining the financial records, payment of debts incurred and maintaining the checkbook." Moreover, the committee suggested a term of five years in order to maintain continuity of detail in this office. The name of a single candidate for Secretary/treasurer, Bill Johansen, was placed on the ballot to be voted on by the membership at the symposium in May, 1982. From this point onwards the monies of NEAS have been in the hands of a single person, except that for several years the DC had a separate account. In the early 1990s, the Secretary/treasurer was given the task of keeping both the main NEAS account as well as the DC account. Moreover, this person was to provide an annual financial report for both of these funds. In 1996, it was recognized that the work load of this person was onerous, and the EC separated the position into those of Secretary and Treasurer. Since 1982, five people have served as Secretary/treasurer or Treasurer of the NEAS: Bill Johansen (1982-1987), Craig Schneider (1987-1992), Bea Robinson (1992-1997), Glen Thursby (1997-2002), and presently, Greg Boyer (2002-2012).

The Society balance carried forward annually in the main account was for many years about $2000, an amount that was just sufficient to begin operations for the next symposium. However, after the 1997 symposium only $244 remained in the NEAS coffers. In part, this was accounted for by low attendance in 1997 (97 registrants), a $700 registration fee charged by MBL and several sizable reimbursements made to invitees. To compensate for the shortfall, registration fees were raised for the 1988 symposium and a late registration penalty was assessed. Recovery to financial stability occurred in one year. During the last few years, the main account has sometimes exceeded $10,000, as has also the DC account. For example, in 2007, the balance in the main account was $10,888, and in the DC account $10,215.

In order to obtain a tax-exempt status, NEAS had to file Articles of Organization and apply for Recognition of Exemption. Largely through the efforts of Secretary/treasurer Bill Johansen the Society finally received tax-exempt status as of 1 July 1987. Members could then deduct dues and donations to NEAS on their individual tax-returns, and income of the Society was no longer taxable. NEAS must file financial reports to the IRS every August. The tax-exempt status necessitated a modification in the NEAS bylaws to reflect the tax code. Each year the Treasurer files with the Commonwealth of Massachusetts a form called the Annual Report. This continues our status as a nonprofit Massachusetts corporation

DISTINGUISHED SPEAKERS

A much-anticipated highlight of each symposium has been the distinguished lecture, or keynote talk. Well-known phycologists are selected at the fall EC meetings and invitations are extended by the co-conveners. In most instances, a single individual is invited, but in a few years two were selected (Table 9). Sometimes the lecture was given after the Saturday banquet, but in recent years it has been scheduled for late Saturday afternoon, the idea being that attendees are likely to be more attentive before cocktails and an ample dinner.

Table 9. *The distinguished lecturers through 2010.*

1976   John Pringle (Halifax, Nova Scotia): *History and development of seaweed utilization in the Canadian Maritimes.*

1977   George F. Papenfuss (University of California, Berkeley CA): *Landmarks in the discovery of sexuality and alternation of generations in the brown algae.*

1978   Frank Trainor (University of Connecticut, Storrs CT): *How do your algae grow?*

1979   John D. Dodge (Royal Holloway College, England): *Dinoflagellates: animal, vegetable and mineral.*

1980   F. J. R. (Max) Taylor (University of British Columbia, Vancouver, Columbia): *The origin of eukaryotic cells.*

1981   Lynn Margulis (Boston University, Boston MA): *Evolution of algae.*

1982   Johan Hellebust (Toronto University, Toronto, Ontario): *Osmoregulation and salt tolerance of algae.*

1983   Lynda Goff (University of California, Santa Cruz CA): *The red algal parasites: what are they trying to tell us?*

1984   Holger Jannasch (WHOI, Woods Hole MA): *Plant life in the deep sea.*

1985   Ursula Goodenough (Washington University, St. Louis MO): *Evolutionary relationships between the sexual agglutinin and the cell wall of* Chlamydomonas.

1986   Mike Levandowsky (Haskins Laboratory, Pace University, New York NY): *Signals and signal processing in the algal protists.*

1987   Matthew Dring (University of Belfast, Northern Ireland): *Throwing light on seaweeds.*

1988   Carole Lembi (Purdue University West Lafayette IN): *'Static' freshwater environments and macrophytic algae: are they compatible?*
        Bob Paine (University of Washington, Seattle, WA): *A zoologists perspective on marine benthic algae as superb material for ecological experimentation.*

1989   Lois Pfiester (University of Oklahoma, Norman OK): *The study of freshwater dinoflagellates: luck and elbow grease.*

1990   Bruce Parker (Virginia Polytechnic Institute and State University, Blacksberg VA): *Origins, evolution and development of phycology in North America.*

1991   Annette Coleman (Brown University, Providence RI): *Species concepts in algae: contributions from molecular biology.*

Robert T. Wilce (University of Massachusetts, Amherst MA): *Role of the Arctic Ocean as a bridge between the Atlantic and Pacific Oceans: fact and hypothesis.*

1992   Don Anderson (Woods Hole Oceanographic Institute, Woods Hole MA): *Toxic dinoflagellate blooms and red tides in New England and abroad: a biogeographic and physiological perspective.*

Frank Trainor (University of Connecticut, Storrs CT): Scenedesmus *phenotypic plasticity: cyclomorphosis.*

1993   Carolyn J. Bird (National Research Council, Halifax, Nova Scotia): *Molecular and other meanderings among the Rhodophyta.*

1994   Linda Graham (University of Wisconsin, Madison WI): *Ecological and evolutionary importance of dissolved organic carbon utilization by charophycean algae.*

James Carlton (Williams College, Williamstown MA and Mystic Seaport, Mystic CT): *Botanical roulette: the potential role of ballast water in the introduction of exotic algae species to North America.*

1995   Bob Guillard (Bigelow Laboratory for Ocean Sciences, West Boothbay Harbor ME): *Natural history and the marine phytoplankton: some directions taken, seen through a case history.*

1996   Bob Anderson (Bigelow Laboratory for Ocean Sciences, West Boothbay Harbor ME): *Algal biodiversity and its significance.*

1997   Robert G. Sheath (University of Guelph, Guelph, Ontario): *Freshwater red algae: from the molecule to the globe.*

1998   Michael Wynne (University of Michigan, Ann Arbor MI): *The complexity of algal systematics: the organisms! The jobs! The future!*

1999   Curt Pueschel (SUNY, Binghamton NY): *Beyond light: fifty years of contributions of electron microscopy to phycology.*

2000   Sarah Gibbs (McGill University, Montreal, Quebec, Canada): *Evolution of peridinin-containing dinoflagellate chloroplasts: surprises and answers.*

2001   Ian Davison (University of Maine, Orono ME): *Travels with PAM: insights into the strange world of seaweed physiology.*

2002   Robert Wilce (University of Massachusetts, Amherst MA): *Musk oxen, fine structure and molecular biology.*

2003   Ralph Quatrano (Washington University, St. Louis MO): *Cytoskeletal and cell wall interactions during the establishment of cell polarity in the* Fucus *zygote.*

2004   Patricia A. Tester (NOAA, National Ocean Service, Beaufort NC): *Copepodology for the phycologist with apologies to G.E. Hutchinson.*

2005   Debashish Bhattacharya (University of Iowa, Iowa City IA): *A phylogenetic and genomic perspective on the origins of photosynthetic eukaryotes within the Tree of Life.*

2006   Graham Underwood (University of Essex, Essex, UK): *Life in estuarine biofilms: small-scale complexity and large-scale importance.*

2007   Dennis Hanisak (Harbor Branch Oceanographic Institution, Fort Pierce FL): *Heroes in the seaweed: John H. Ryther.*

2008   Michael Guiry (Martin Ryan Institute, NUI, Galway, Ireland): *AlgaeBase.*

2009   Scott Gordon (Green Technologies, Winooski VT): *Small scale biodiesel production (including some do's and don't's for making biodiesel from algae).*

2010   Olivier De Clerck (Universiteit Gent, Gent, Belgium): *Species, patterns and algae: a phylogenetic perspective* (presented via internet connection due to flight restrictions).

AWARDS

A natural event at symposia of the type organized by NEAS has been the presentation of awards to individuals who in one way or another have done outstanding work on behalf of phycology or NEAS. The first types of awards presented by the Society were to students, an indication of the importance that NEAS places on young, developing scientists. In recent years, some student award winners have received one-year subscriptions to the Journal of Phycology, thanks to the PSA.

Student awards. In keeping with the goals of NEAS, various types of awards to students are regularly given out at each symposium. Foremost among these awards are those for the best presentations by graduate students, both for their talks and for their posters. Judging committees are appointed by co-conveners for selecting the recipients of awards. The EC, in 1989, decided to name these awards the 'Robert T. Wilce Student Awards' beginning at the next symposium in 1990 (Table 10). This action was in recognition of the enormous contributions Wilce has made to the Society; he was surprised by the honor at the Saturday evening banquet. Each award constituted a document on archival paper (Figure 29) as well as a check for $75. These awards are made from the general fund, not from the DC account.

In April, 2009, the EC voted, after hearing the strong support of Bob Wilce, to rededicate the Wilce poster award. Thus, beginning with the symposium in 2010, the graduate student selected as having the best poster would receive the 'Frank R. Trainor Award.' As may be imagined, Trainor was pleased to hear of this honor, a well deserved one.

Recognizing that more and more undergraduate students were attending the symposia the EC decided, in 1998, to begin awarding a monetary prize of $50 for the best undergraduate presentation, and calling it the 'President's Award.' Funds were to come from the general NEAS account, not from the DC. The first President's Award was made at the symposium in 1999, and the practice has been perpetuated (Table 11).

The DC made its first travel award to a graduate student in 1988. The next year, in 1989, monetary awards were made to four students in an amount totaling $335. In general, the awards given out have been increasing in number and value year by year as the DC has been successful in soliciting funds. In 1995, 22 student awards were made, most of them for travel to the symposium.

Certificate of Recognition
from
The Northeast Algal Society
The Robert T. Wilce Award
for the meritorious paper presentation
by a student at the   Northeast Algal Symposium.

Marine Biological Laboratory, Woods Hole, Massachusetts
April, 19

CHAIRMAN, NEAS

SEC. • • NEAS

Figure 29   The Robert T. Wilce award for the best student oral presentation. The Poster award is similar.

For several years, the DC has been giving out book awards to students at the symposium. The rules were that to win a book a student should be attending the symposium and, in a write-up, explain how the book will enhance his/her education.

In 1996, NEAS sponsored a photo contest, with 8x10-inch macro and micrographs of algal subjects. Barry Colt and Bill Johansen solicited photographs and organized judging committees. First and second place finishers in each category were given monetary awards and the photos themselves were deposited with the DC for possible future incorporation into a phycological calendar. The contest was held in 1996 and 1997, and then discontinued.

In addition to the awards provided by NEAS several commercial firms have provided awards, such as books, microscopes, and the like, that have been given to students at the banquet ceremonies following the meal. Selection of the recipients has been by random drawings. Foremost among the firms doing this have been Connecticut Valley Biological Supply and the Swift Microscope Company.

Table 10. *Graduate student award winners from the first year awards were given in 1983 until the present. Beginning in 1990 both the award for the best talk as well as that for the best poster were designated the Robert T. Wilce Awards. Beginning in 2010 the Wilce poster award was renamed the Frank R. Trainor Award.*

1983	Al Steinman, University of Rhode Island
	John Hackney, Georgetown University
1984	Donna Johnson, University of Connecticut
	Rod Fujita, MBL
1985	Sara M. Lewis, Harvard University
	Martha Ludwig, McGill University
1986	Juan Correa, Acadia University, Wolfville, Nova Scotia
1987	Rick Greene, SUNY at Stony Brook
1988	Steve Dudgeon, University of Maine
1989	R.C. Sokol, SUNY at Albany
1990	Janet E. Kuebler, University of Maine
	Josh P. Philibert, University of Massachusetts
1991	Julie Yates, Smith College
	Katherine Duff, Queen's University
1992	Steve Wilhelm, University of Western Ontario
	Kimberly Brown, Queen's University
1993	Y. H. Zhou, National Research Council
	Anne-Marie Lott, Connecticut College
	Kelly Murphy, University of Western Ontario
1994	Andrea Nerozzi, Brown University
	Christopher Harley, Brown University
1995	Unknown
1996	Keith Josef, Syracuse University
	Susan J. Babuka, SUNY at Binghamton

1997	Unknown
1998	Douglas McNaught, University of Maine
	Todd Harper, University of New Brunswick
	Bradley Metz, Northeastern University
1999	Robert Verb, Ohio University
	Sean P. Grace, University of Rhode Island
2000	M. Lynn Berndt, University of Maine
	Jennifer Dalen, University of New Brunswick
2001	Penny Humby, University of New Brunswick
	Xingye Yang, SUNY at Syracuse
2002	Colin R. Bates, University of New Brunswick
	Aaron Wallace, University of New Hampshire
2003	Brian W. Teasdale. University of New Hampshire
	Susan L. Clayden, University of New Brunswick
2004	Andrew Bamburger, Great Lakes Institute of Ecology
	Karen Pelletreau, University of Delaware
2005	Priya Sampeth-Wiley, University of New Hampshire
	Jeremiah Hackett, University of Iowa
	Lisa Pickell, University of Maine
2006	Hilary A. McManus, University of Connecticut
	Daniel McDevit, University of New Brunswick
2007	Jessica Muhlin, Maine Maritime Academy
	Malcolm McFarland, University of Rhode Island
2008	Nathan J. Smucker, Ohio University
	Yen-Chun Liu, Northeastern University
2009	Susan L. Clayden, University of New Brunswick
	Bridgette Clarkson, University of New Brunswick
2010	Molly Letsch, University of Connecticut (Wilce Award)
	Jeremy Nettleton, University of New Hampshire (Trainor Award)

Table 11. *Recipients of the President's Award for the best undergraduate student oral or poster presentation.*

1999	Amy E. Cocina, SUNY at Geneseo
2000	Tomas A. Bonome, Colgate University
	Andrea Shelley, SUNY at Geneseo (runner-up)
2001	Julie Price, University of New Brunswick
	Jennifer Sallee, Roger Williams University
2002	Anthony Gallo, SUNY at Geneseo
2003	Hannah A. Shayler, Connecticut College
	Martin P. Monahan, Eastern Connecticut College
2004	David Sakoda, Wheaton College
	Megan Brennan, Susquehanna University
2005	Katie Williams, University of New Hampshire
2006	Cayelan Carey, Dartmouth College
2007	Laura Lambiase, Skidmore College
2008	Sara Hall, University of Maine
	Jonathon Neilson, University of New Brunswick
2009	Elizabeth Sargent, Roger Williams University
2010	Elisabeth Cianciola, Trinity College

The Frank Shipley Collins Award. In 1996, the Frank Shipley Collins award for meritorious service to the NEAS and to phycology in general was inaugurated with the recognition of Edwin Boger of Worcester State College. The award was initiated by the EC to honor professional phycologists who had supported NEAS above and beyond the call of duty (Table 12).

The award commemorates Frank S. Collins (1848-1920), by vocation an accountant for the Boston Rubber Shoe Company in Malden, Massachusetts, and by avocation, an ardent collector and amateur phycologist. Throughout his career, Collins wrote several books and published more than 100 papers on both marine and freshwater algae primarily from New England, but also from Pacific North America, the Caribbean, the Bahamas and Bermuda. Another great legacy of Collins was the innumerable specimens he left in a wide range of herbaria in North America and Europe, including his monumental contribution to the most extensive published exsiccate of North American species in *Phycotheca Boreali-Americana* (Collins, Holden & Setchell, 1895-1919). The preparation of this exsiccate near the turn of the 20th century, 51 fascicles of mostly 50 specimens for each of 80 sets, was a labor of love that fell primarily to Collins himself. All of these accomplishments were completed in the lifetime of a man with only a high school diploma and a daytime accounting job, "above and beyond the call of duty" for any naturalist of his time (Moe and Browne, 1996).

Table 12. *Recipients of the Frank Shipley Collins Award, 1996-2010.*

Ed Boger, 1996: a strongly positive force on the DC from its inception in 1984; chaired the DC from 2002 to 2003; innovative ideas and exceptional work ethic were instrumental in the success of the DC for many years; chair of the Nominations Committee in 1988.

Barry Colt, 1997: co-convener in 1993; chair of the DC for several years; auctioneer at NEAS post-banquet auctions in the mid-1990s; designed logo in 1993.

Bill Johansen, 1998: co-convener in 1978 and 2009; Secretary/treasurer 1982-1987; NEAS President 1987-1992; chaired Nominations Committee in 1993 and 1998; membership director and archivist 1995-2000; member of the Publications Committee.

Bea Robinson, 1999: a much-loved regular attendee from the 1970s until she died of cancer early in 2000; Secretary/treasurer from 1992 to 1997.

Jim Sears, 2000: co-convener in 1969 and 1979, and in 1991; member-at-large 1986-1989; Chaired Publications Committee for several years; edited '*NEAS Keys to Benthic Marine Algae of the Northeastern Coast of North America from Long Island Sound to the Strait of Belle Isle*' in 1998 and 2002 (Fig. 27).

Bob Wilce, 2001: co-convener in 1969, 1976, and 2009; first NEAS President from 1982-1985; member of the ad hoc committee of 1981-1982; created the first draft of the NEAS bylaws; member-at-large 1997-2000; Honorary chairperson and Distinguished lecturer in 1991; Distinguished lecturer 2002; member of the Publications Committee (Fig. 26).

Frank Trainor, 2002: co-convened the first NEAS symposium outside the New York metropolitan area in 1969, co-convened again in 1973 and 1977; Distinguished lecturer in 1978; Honorary chairperson and Distinguished lecturer (again) in 1992; Honorary chairperson in 2009 (Fig. 28).

Craig Schneider, 2003: co-convened in 1980 and 2009; led field trips; Secretary/treasurer 1987-1992; NEAS President 1996-2000; first chair of the DC; post-banquet auctioneer; invited lecturer in 1991 and 2010.

Not awarded, 2004.

Larry Liddle, 2005: enthusiastic attendee from Long Island University, often bringing students; co-convener in 1986; NEAS President 1992-1994.

Curt Pueschel, 2006: attends all symposia; distinguished lecturer in 1999; NEAS President 2000-2004.

Gary Saunders, 2007: most evident Canadian phycologist attending the NEAS symposia bringing many students; convener 1999 and 2003; invited lecturer in 2009; NEAS President 2004-2006.

Glen Thursby, 2008: celebrated NEAS auctioneer; active on DC; co-convener 1997 and 2007; Treasurer 1997-2002.

Peter Bradley, 2009: regular symposium attendee with wife Rosslyn; took over as Secretary in 1996, and re-elected in 2001-2006; stickler for detailed minutes.

Greg Boyer, 2010: co-convener in 1996; NEAS treasurer for 10 years 2002-2012; very careful with NEAS monies.

HONORARY CHAIRPERSONS

The idea of honoring senior phycologists at the annual symposia had its genesis in a letter from Frank Trainor to Bob Wilce in May, 1983. In the letter, Trainor presented a motion: "That we have an honorary chairperson for our Northeast meetings whenever the occasion is right." He further added that the chosen persons should be "senior phycologists." When brought to the EC meeting in October, 1983, Trainor's suggestion received a warm reception. Discussion of this prospect brought several possible candidates to mind, and it was decided to ask William Randolph Taylor to be our first NEAS honorary chairperson at the symposium in 1984.

Wm. R. Taylor (1984). Anyone who has carried out taxonomic or biogeographic work on seaweeds in the western Atlantic, the Caribbean and Pacific Mexico, has consulted the books and papers of William Randolph Taylor (1895-1990). From the late 1920s until 1989 Taylor meticulously recorded floristic information from his many field trips and from the literature. The selection of Taylor as the first NEAS honorary chairperson was an easy choice. On top of his prodigious floristic work, Taylor had a long-standing connection with MBL (from 1917-1989) and a home in Woods Hole where he and his wife Jean visited most summers (Wynne 1996). It was therefore very fitting that NEAS, with much of the legwork done by Wilce, should honor him by soliciting funds for a Lillie Chair that could be named in his behalf. A Lillie Chair adorned with a brass engraved plate in the Lillie Auditorium cost $1000 in 1984 and it is tribute to Taylor that far more than that amount was collected.

Hannah T. Croasdale (1985). Croasdale (1905-1999) was a remarkable person, as attested to in the year she was NEAS honorary chairperson. Like her Ph.D. mentor Wm Randolph Taylor, she was a formidable part of the MBL marine botany program starting in 1928 or 1929. Many commendable adjectives have been used in describing Hannah Croasdale, with friendly, energetic, indomitable and knowledgeable among the most common. She was one of the founders of PSA. She was the first woman to work her way up through the ranks to become a full professor (in 1968) at Dartmouth College, after beginning as a Technical Assistant in 1935. She retired in 1971, but kept on with most of her scholarly activities. Her primary research interest was desmids, and soon after 1935 she was recognized as a world authority on these algae. Croasdale's scholarly interests also included Latin, and many authors of new taxa owe a huge debt to Croasdale for converting diagnoses from English into Latin. Croasdale was also a top-notch illustrator. But, perhaps most of all she was an indefatigable and inspiring teacher, and many students accompanying Croasdale on field trips have fond memories of what they learned and, incidentally, of how they had difficulty in keeping up with their teacher. This, even after she had retired and had two artificial hips.

Luigi Provasoli (1986). Provasoli (1908-1992) led a rich and varied life in scientific research on insects, and then protists. He was Honorary Chair of our 25th symposium (1986) where he was well remembered for his success in developing culture media in which to grow various phytoplankters. At that meeting we all learned about some of Luigi's personal attributes. He was born in Italy in 1908 and died in Italy in 1992. After coming to the United States with his wife, Rose, he became intensely involved in research at Haskins Laboratories, first located in New York and then at Yale University. He retired in 1987. In addition to research he served on Presidential Science Advisory Committees, was President of the PSA and founding editor of the *Journal of Phycology*.

Among his many extra duties he also served on the boards of the American Institute of Biological Sciences and the American Type Culture Collection. A memorial tribute written by Lehman and Lehman (1996) tells that Luigi Provasoli was a very warm, sincere and friendly gentleman.

Edwin Theodore Moul (1988). Moul (1903-1988) passed away just before he would have served as Honorary chair of the 27th NEAS symposium. He is remembered for his studies of mosses in Pennsylvania and plant and invertebrate animal surveys along the coasts of southern New England and the estuaries of the mid-Atlantic coast. He held positions at Rutgers University and at the Woods Hole Oceanographic Institute. Ed Moul was a floristics scientist and an expert on the biology of the seashores of Cape Cod. After Moul retired from Rutgers, he became deeply involved in documenting the effects of oil spills on Cape Cod salt marshes. In the 1970s, he worked with oceanographer George R. Hampson of WHOI to report on their three-year Cape Cod pollution study in the Fisheries Research Board of Canada. Moul and Hampson's pioneering oil pollution studies were carried on and expanded by others well into the 1990s. But Moul also ranged widely, publishing reports, for example, on the flora and land fauna of Onotoa Atoll in the Gilbert Islands. In the 1970s, he regularly opened the doors of his Woods Hole herbarium to visitors attending the annual NEAS symposia.

Bud Brinkhuis (1990). Brinkhuis (1946-1989) was a regular and active attendee at NEAS meetings until his sudden death in the summer of 1989. His passing was additionally sad for all of us at NEAS because he was about to begin his term as chairperson, a post to which he had been elected at the symposium in April of that year. Earlier (in 1983), he and Peter Heywood had convened the twenty-second annual symposium. For several years, Bud maintained the NEAS mailing list. Bud, an Assistant Professor in Stony Brook's Marine Sciences Research Center, had been very active in phycology. His research dealt with the ecophysiology of seaweeds (Yarish, 1988).

Robert T. Wilce (1991). Wilce (Figure 30), now a Professor Emeritus at the University of Massachusetts, has devoted much of his exuberant personality to NEAS, beginning in the late 1960s (he was a co-convener in October 1969) and continuing to the present. A list of his major roles in NEAS is given in Table 12, which is based on information procured when he was awarded the Frank Shipley Collins Award in 2001. Bob has had a rich and varied life. During World War II he was parachuted into the Battle of the Bulge with the 17th Airborne Division. He was 100% disabled from wounds received near Bastogne, Belgium, in January, 1945, shortly after the Battle of the Bulge. Bob's formal academic education came in the 1950s, topped off by a Ph.D. from the University of Michigan, Ann Arbor, under the mentorship of Wm Randolph Taylor. Much of Bob's research has been on taxonomic and biogeographic questions concerning arctic and subarctic seaweeds. He takes pride in having lived by frigid arctic waters for many months at a time and for SCUBA diving in these waters, all in the interests of algae. In 1998, the University of Copenhagen recognized Wilce's voluminous contributions to arctic marine biology when it awarded him an Honorary Doctor of Science degree.

Figure 30. Bob Wilce.

Figure 31. Frank Trainor

Francis Rice Trainor (1992, 2009). Trainor (Figure 31) has been an institution at the University of Connecticut's Storrs campus since he joined its faculty in 1957. He earned both his masters (1953) and doctorate (1957) degrees at Vanderbilt University. Frank has been prominent in NEAS ever since he went to the sixth symposium in 1968. A year later, he and Joanna Page organized the seventh symposium in Storrs. He then convened the twelfth symposium in 1973 and the sixteenth (with Joe Ramus and Sam Beale) in 1977. Frank is an idea man; for example, he suggested in 1983 that NEAS have an occasional honorary chair at its symposia. He also has a contagious sense of humor, something all attendees witnessed at his NEAS distinguished lecture entitled "*How do your algae grow?*" at the seventeenth symposium (1978). Incidentally, he also presented a distinguished lecture at the thirty-first symposium in 1992 when he was Honorary chair. Frank also has a very creditable research record with many publications, a text on algae, and a book-length volume on *Scenedesmus* phenotypic plasticity. In 1970, he was a Fulbright Research Scholar in Stockholm and in 1971 a Fulbright Lecturer in Greece and Yugoslavia. He was president of the PSA in 1969 and won the Darbaker Award from the Botanical Society of America in 1965. With the green algal genus *Scenedesmus* as his main research interest he has worked extensively on sexual reproduction, algal nutrition, and morphogenesis. He retired in 1997 but seems almost as active in phycology and NEAS as ever.

Annette W. Coleman (1993). Coleman received her A.B. at Barnard College in 1955 and her Ph.D. in 1958 at the University of Indiana under the direction of Richard Starr. Here she developed a career-long interest in volvocalian colonies. Before taking a position at Brown University in 1963, where she is currently the Stephen T. Olney Professor of Natural History (since 1985), she spent three years as a NSF Research Fellow at Johns Hopkins University and a research associate at the University of Connecticut. Among her many honors she is a two-time recipient of the Provasoli Award for the best paper in the *Journal of Phycology* (1983, 1988). In 1986, she received the Botanical Society of America Darbaker Award and in 1990, she became a Fellow of the American Association for the Advancement of Science. She was president of the PSA (1981-82) and was the second president of NEAS (1985-1987). Annette's main research interests are mating physiology in Volvocales, chloroplast DNA and gaining further data on cell fusion. Her extensive collaboration with Linda Goff has dealt with fluorochrome analyses of DNA in red algae and other issues.

Sarah P. Gibbs (2000). Gibbs, Macdonald Emeritus Professor of Botany at McGill University, obtained B.A. and M.A. degrees at Cornell University and then her Ph.D. in cell biology at Harvard University in 1962. She has published epic papers on algal plastids with a focus on understanding the evolution and molecular organization of algae. In some of her most exciting work, she and her students showed that some algae in the distant past engulfed and incorporated cells that became endosymbionts serving as plastids. The evidence is in the four plastid membranes, the extra ones being vestigial cell membranes of the endosymbionts. Recently, her research team has been studying the separation and evolution of photosystems I and II, a problem that has a bearing on the evolution of vascular plants from certain algae. Her research, as well as her teaching, has resulted in several honors, with her 1999 Award of Excellence from the Phycological Society of America among her most treasured. She has also received the Darbaker Prize from the Botanical Society of America and she is a Fellow of The Royal Society of Canada and The American Association for the Advancement of Science.

Arthur C. Mathieson (2004). Mathieson earned his doctorate at the University of British Columbia in 1965, and is still active in 'systematics, floristic, and autecological studies of seaweeds, particularly introduced and economically important taxa.' He has been at the University of New Hampshire since 1965, and from 1972-1982 he was director of the Jackson Estuarine Laboratory. Art has written hundreds of scientific papers and recently (2008) collaborated with C.J. Dawes to write '*The Seaweeds of Florida*,' a tome of almost 600 pages. Almost from the beginning Art has played a role in NEAS. He was involved in the seventh symposium, in 1969. He co-convened in 1981, was an invited speaker in 1983, and was our NEAS Honorary chair in 2004.

Robert Vadas (2005). Robert L. Vadas, Sr. (Bob) began his long-time interest and extensive research on marine and algal ecology when he worked on his doctorate at the University of Washington. He earned his PhD in 1968. His long career at the University of Maine has included many awards and honors, and, in spite of health problems, he is still active at that institution. In 1989 he received a 'Visiting Senior Scientist Award' for research in Norway. His many publications have included ones co-authored with Bob Paine, Bob Steneck, and Ian Davison, all of whom have been affiliated with NEAS in one way or another. A seminal paper with Paine in 1969 on benthic algal populations and sea urchin grazing has been cited hundreds of times. He has even co-authored a short scientific note with his son, Robert L. Vadas, Jr. (1995). His interests have included NEAS, where he gave a special lecture in 1981, was a co-convener in 1988, contributed to the NEAS publication of the keys to benthic marine algae edited by Jim Sears, and, finally, Bob was the NEAS Honorary Chair in 2005.

H. William Johansen (2006). Johansen began his work on algae in 1962 under George F. Papenfuss (NEAS distinguished lecturer, 1977). After a doctorate in 1966 at UC Berkeley, Bill spent a post-doc becoming familiar with the amazing diversity of coralline algae on the Indian Ocean shores of South Africa. In 1968, he took a position at Clark University. His research dealt mostly with coralline algae, and included a book on the subject in 1981. His teaching included more than 15 trips with student groups to Bermuda and three one-month teaching sessions in a program in Luxembourg. During his 31 years at Clark University, he took a great interest in NEAS, becoming a co-convener in 1978, Secretary/treasurer 1982-1987, President 1987-1992, and Honorary chair in 2006. During his retirement he has been writing a book about an Italian immigrant community that developed a century ago in a little Yankee village in Vermont.

Table 13. *Honorary NEAS chairpersons.*

1984 William Randolph Taylor, University of Michigan
1985 Hannah Croasdale, Dartmouth College, New Hampshire
1986 Luigi Provasoli, Haskins Laboratories, New York
1988 *Ed Moul, Rutgers University, New Jersey
1990 *Bud Brinkhuis (Boudewijn), SUNY Stony Brook
1991 Bob Wilce, University of Massachusetts
1992 Frank Trainor, University of Connecticut
1993 Annette Coleman, Brown University
2000 Sarah P. Gibbs, McGill University
2004 Art Mathieson, University of New Hampshire
2005 Bob Vadas, University of Maine
2006 Bill Johansen, Clark University
2009 Frank Trainor, University of Connecticut

*Deceased prior to the symposium.

CONCLUSIONS

After a tentative beginning in 1966, what we now know as the Northeast Algal Society (NEAS) has grown into a formidable society for students and scientists interested in marine and fresh water algae in New England and eastern Canada, as well as in many other parts of the world. The success of NEAS is due to its focus on students (both undergraduates and graduates), while at the same time recognizing professional phycologists in annual symposia characteristically attended by a unique and satisfying mix of personalities.

Our goals in this document are to detail the significant people and events that have led to NEAS as it now operates. For most of the existence of NEAS, the symposia have been held on spring weekends (usually in April), typically drawing between 100 and 200 attendees. Most NEAS symposia have been organized primarily by a pair of scientists, the co-conveners. Since 1982, the co-conveners have been assisted by an EC. At that time NEAS also became affiliated with PSA.

From 1976 until 1997, the annual symposia were held at the Marine Biological Laboratory (MBL) in Woods Hole, a very inspiring setting for phycologists. However, since 1997, NEAS has convened at convention centers elsewhere in New England. In April, 2011, after a hiatus of fourteen years, the fiftieth symposium will again convene at MBL in Woods Hole.

Following the lead of NEAS, two other regional organizations were established, the Southeastern Phycological Colloquy in 1978 and the Northwest Algal Symposium (NWAS) in 1984. The southern organization meets annually in a less formal way than NEAS. In recent years the NWAS has met annually, the last time in April, 2010, at Whidbey Island, Washington. According to their website they draw from 60-120 people per symposium.

ACKNOWLEDGMENTS

Many, many individuals, especially those in NEAS Executive Committees through the years, have helped in preparing this document, and we thank them sincerely. We are especially grateful to the following, who provided us with much information and proof-read drafts of either the current manuscript or the previous one (2003): Sheldon Aaronson, Carolyn Bird, Ed Boger, Susan Brawley, Annette Coleman, Barry Colt, Norm Lazaroff, Louise Lewis, Larry Liddle, Curt Pueschel, Craig Schneider, Jim Sears, Frank Trainor, Peg Van Patten, and Bob Wilce. Wilce and Schneider proofread the entire manuscript and offered many valuable suggestions. Van Patten, chair of the current NEAS Publications Committee, and Connecticut Sea Grant were a tremendous help in the final stages of completing the document. It has been a great pleasure working with them.

REFERENCES

Johansen, H.W. & P.M. Bradley. 2003. *A History of the Northeast Algal Society (NEAS).* NEAS Contribution Number 3. 58 pp.

Lehman, J.T. & D.A. Lehman. 1996. Luigi Provasoli (1908-1992). Pp. 327-336. In Garbary, D.J. and M.J. Wynne (eds.). *Prominent Phycologists of the 20th Century.* Lancelot Press Limited, Hantsport, Nova Scotia. 360 pp.

Maggs, C. 1998. Book review of Sears 1998. *Journal of Phycology* 34(6):1094-95.

Moe, R.L. & D. Browne. 1996. W.A. Setchell (1864-1943 and N.L. Gardner (1864-1937). Pp. 102-114. In Garbary, D. J. and M. J. Wynne (eds.). *Prominent Phycologists of the 20th Century.* Lancelot Press Limited, Hantsport, Nova Scotia. 360 pp.

Sears, J.R. (ed.). 1998. *NEAS Keys to Benthic Marine Algae of the Northeastern Coast of North America from Long Island Sound to the Strait of Belle Isle.* NEAS Contribution Number 1. 161 pp.

Sears, J.R. (ed.). 2002. *NEAS Keys to Benthic Marine Algae of the Northeastern Coast of North America from Long Island Sound to the Strait of Belle Isle* (Second Edition). NEAS Contribution Number 2. 161 pp.

Wynne, M.J. 1996. William Randolph Taylor (1895-1990). Pp. 175-183. In Garbary, D.J. and M. J. Wynne (eds.). *Prominent Phycologists of the 20th Century.* Lancelot Press Limited, Hantsport, Nova Scotia. 360 pp.

Yarish, C. 1988. Boudewijn H. Brinkhuis 1946-1989. *Applied Phycology Forum* 6(2): 9.

APPENDICES

Appendix A. *The co-conveners and locations of NEAS symposia (see Appendix D for full names and affiliations).*

1   1966 (Spring): Sheldon Aaronson and Bill Siegelman, at Queens College, Flushing NY
2   1966 (Fall): Bill Siegelman and Harvard Lyman, at Brookhaven National Laboratory, Upton NY
3   1967 (Spring): Ray Jones, at SUNY at Stony Brook NY
4   1967 (Fall): Ruth Sager and Marcia Brody, at Hunter College, New York NY
5   1968 (Spring): ?? at Yale University, New Haven CT
6   1968 (Fall): Melvyn I. Selsky and Melvin M. Belsky, at Brooklyn College NY
7   1969 (Spring): Joanna Z. Page and Frank Trainor, at the University of Connecticut, Storrs CT
8   1969 (Fall): Jim Sears and Bob Wilce, at the University of Massachusetts, Amherst MA
9   1970 (Fall): Norm Lazaroff, at SUNY at Binghamton NY
10 1971————————————-no meeting?————————————-
11 1972————————————-no meeting?————————————-
12 1973 (Apr): Frank Trainor, at the University of Connecticut, Storrs CT
13 1974————————————-no meeting?————————————-
14 1975: Roger Goos and Marilyn Harlin, at the University of Rhode Island, Kingston RI
15 1976: Ray Jones and Bob Wilce, at MBL, Woods Hole MA
16 1977: Sam Beale, Joe Ramus, and Frank Trainor, at Woods Hole
17 1978: Steve Golubic and Bill Johansen, at Woods Hole
18 1979: Annette Coleman and Jim Sears, at Woods Hole
19 1980: Phil Cook and Craig Schneider, at Woods Hole
20 1981: Nina Allen and Art Mathieson, at Woods Hole
21 1982: Beth Gantt, Hank Parker, and Charlie Yarish, at Woods Hole
22 1983: Bud Brinkhuis and Peter Heywood, at Woods Hole
23 1984: Don Cheney and Llewellya Hillis-Colinvaux, at Woods Hole
24 1985: Susan Brawley and Robert Sheath, at Woods Hole
25 1986: Larry Liddle and Bea Robinson, at Woods Hole
26 1987: Paul Hargraves and Tom Lee, at Woods Hole
27 1988: Phil Sze and Bob Vadas, at Woods Hole
28 1989: Bob Steneck and John Van der Meer, at Woods Hole
29 1990: Ruth Schmitter and Pete Siver, at Woods Hole
30 1991: Dick Fralick and Jim Sears, at Woods Hole
31 1992: Mike Levandowsky and Aimlee Laderman, at Woods Hole
32 1993: Barry Colt and Hank Parker, at Woods Hole
33 1994: Paulette Peckol and Curt Pueschel, at Woods Hole

34 1995: Dorothy Swift and Gary Wikfors, at Woods Hole

35 1996: Carolyn Bird and Greg Boyer, at Woods Hole

36 1997: Glen Thursby and John Wehr, at Woods Hole

37 1998: Tracy Villareal and Joby Chesnick, at the Plymouth Sheraton Inn MA

38 1999: Peg van Patten and Gary Saunders, at the Plymouth Sheraton Inn MA

39 2000: Harold Hoops and Morgan Vis, at the Whispering Pines Conference Center, W. Alton Jones Conference Campus RI

40 2001: Karen Culver-Rymsza and Carl Grobe, at the Plymouth Sheraton Inn MA

41 2002: Anita Klein, Jack Holt, and Chris Neefus, at the University of New Hampshire NH

42 2003: David Domozych and Gary Saunders, at Skidmore College, Saratoga Springs NY

43 2004: Louise Lewis, Charles Yarish, and Senjie Lin, at Avery Point, University of Connecticut, Groton CT

44 2005: Bob Anderson and Charles O'Kelly, at Samoset Resort, Rockland ME

45 2006: Ray Kepner, John Heimke and David Domozych, at Marist College, Poughkeepsie NY

46 2007: Glen Thursby and Morgan Vis, at Village Inn, Narragansett RI

47 2008: Chris Neefus and Anita Klein, at the University of New Hampshire, Durham NH

48 2009: Bob Wilce, Craig Schneider, and Bill Johansen, at the University of Massachusetts, Amherst MA

49 2010: Brian Wysor and Chris Lane, at Roger Williams University, Bristol RI

50 2011: Louise Lewis and Gary Saunders, at MBL, Woods Hole MA

Appendix B. *NEAS Bylaws as of October, 2007.*

NORTHEAST ALGAL SOCIETY, INC.
BY-LAWS

ARTICLE I: Name.

The name of this organization shall be the Northeast Algal Society, Inc. (NEAS). NEAS is a separately organized nonprofit Massachusetts corporation and a section of the Phycological Society of America.

ARTICLE II: Purposes.

Said Society is organized exclusively for scientific and educational purposes within the meaning of Section 501 (c) (3) of the Internal Revenue Code and shall remain such an organization without object of financial gain. Its purposes are to encourage and promote the growth and development of phycology as a pure and applied phase of biological sciences and to promote the general welfare and good fellowship of phycologists and other biologists, particularly in the States and Provinces of the geographical region of northeastern North America. No part of the net earnings shall inure to the benefit of private shareholders or other members of the Society.

ARTICLE III: Membership.

Section 1. The Northeast Algal Society shall include dues-paying members who may be any member of the general public in sympathy with the purposes of the Society.

Section 2. Members of the Society shall pay annual dues in an amount determined by the Executive Committee. There are two categories of dues-paying members: professionals and students. Student membership shall be limited to seven years.

Section 3. Only members of the Society shall be entitled to vote or hold office.

Section 4. No member of the Society shall be entitled to any distributive share of its assets and, in the event of dissolution, the assets remaining after payment of its debts shall be distributed, as determined by majority vote of the members of the Society, to an organization meeting the conditions prescribed, at the time of dissolution, by Section 501 (c) (3) of the Internal Revenue Code.

ARTICLE IV: Meetings.

Section 1. There shall be one annual Symposium, during which time a business meeting will be held.

Section 2. The time and place of the annual Symposium, business meeting, and any other meetings, shall be determined by the Executive Committee.

Section 3. The members in attendance at the business meeting of the Society shall constitute a quorum to transact any necessary business.

ARTICLE V: Executive Committee.

Section 1. The Executive Committee of the Northeast Algal Society shall consist of a President; a Vice President/President-elect (in alternate years); the two Conveners; the two Conveners-elect, to assume the duties of the Conveners in the following year; the two Conveners from the prior year; a Secretary; a Treasurer (serving as Clerk); two Members-at-large; a Membership Director; and such non-voting members as those specified in Article VI, Section 5. The President shall serve for two years. The Vice President/President-elect shall serve for one year before assuming the office of President. The Membership Director, Secretary, and Treasurer shall serve for five years. The Conveners shall serve for three years. Members-at-large shall serve for three years in staggered terms.

Section 2. Officers shall be eligible for re-election.

ARTICLE VI: Committees and Student Representation.

Section 1. The Executive Committee shall be a standing committee. This committee may replace any of its members, including the President, owing to the inability of a member to complete a term in office. Tenure of a replacement member of the Executive Committee shall terminate at the time of the annual Symposium.

Section 2. The Nominations Committee shall consist of a Chair, elected by ballot at the previous Symposium, and two additional members who will be appointed by the Chair with review by the President.

Section 3. The Development Committee shall be a standing committee consisting of no less than three members. Members are appointed by the President in consultation with the committee chair. Terms shall begin upon appointment and continue for three full calendar years. Members shall be eligible for reappointment.

Section 4. The Publications Committee shall be a standing committee consisting of no less than three members. Members are appointed by the President in consultation with the committee chair. Terms shall begin upon appointment and continue for three full calendar years. Members shall be eligible for reappointment.

Section 5. The President may establish and appoint *ad hoc* committees on an annual or biannual basis to fulfill the goals of the Society.

Section 6. Student representatives shall be the winners of the Robert T. Wilce Awards from the contributed poster and paper sessions at the last annual Symposium of the Society.

Section 7. The chairs of all ad hoc committees and the Student Representatives shall be non-voting members of the Executive Committee.

ARTICLE VII: Election of Officers.

Section 1. The annual election shall be conducted by the three-person Nominations Committee.

Section 2. A slate of candidates will be provided by the Nominations Committee. The election of officers by the members of the Society shall be by mail or at the annual Symposium, with opportunity provided for write-in candidates. Terms of office shall begin and end at these Symposia except for the Treasurer's term. The Treasurer's term of office shall end prior to the Fall Executive Committee Meeting after business from the Annual Symposium has been taken care of and at that time the new Treasurer shall take office.

Section 3. The Executive Committee shall elect a Membership Director who will be a voting member of the Executive Committee.

ARTICLE VIII: Amendments.

Proposed amendments to the Bylaws of the Northeast Algal Society must be submitted in writing to the Secretary. Such proposals shall be reviewed and approved by the Executive Committee before being put to vote by the membership. A two-thirds majority of voting members shall be required to ratify any proposed amendment.

ARTICLE IX: Nonprofit Status.

No part of the assets of the Corporation and no part of any net earnings of the Corporation shall be divided among or inure to the benefit of any officer or director of the Corporation or any private individual or be appropriated for any purposes other than the purposes of the Corporation as herein set forth; and no substantial part of the activities of the Corporation shall be the carrying on of propaganda, or otherwise attempting to influence legislation, and the Corporation shall not participate in, or intervene in (including the publishing or distributing of statements), any political campaign on behalf of any candidate for public office. It is intended that the Corporation shall be entitled to exemption from federal income tax under Section 501 (c) (3) of the Internal Revenue Code and shall not be a private foundation under Section 509 (a) of the Internal Revenue Code.

ARTICLE X: Dissolution.

Upon the liquidation or dissolution of the Corporation, after payment of all of the liabilities of the Corporation or due provisions therefore, all of the assets of the Corporation shall be disposed of to one or more organizations exempt from federal income tax under Section 501 (c) (3) of the Internal Revenue Code (or the corresponding provision of any future United States Internal Revenue Law).

Appendix C. *NEAS Officers Manual, as of 1 February 2011.*

President

1. Call and preside at the fall meeting of the Executive Committee at a site and date convenient to as many members as possible. This meeting is generally held at the site of the upcoming annual symposium to allow members of the Executive Committee to preview and comment on the site.
2. Appoint ad hoc committees to carry out the Society's business outside the scope of standing committees.
3. Consult with committee chairs and the Membership Chair on the appointment of members of standing and ad hoc committees, including vacancies needing members.
4. Preside at the Executive Committee meeting, general business meeting, and the banquet at the annual symposium.
5. Use the office to advance the goals of the Society.
6. Monitor, with sensitivity, other officers' activities in executing their obligations to the Society in a timely fashion.
7. Verify arrangements for the annual symposia with co-conveners at least a month before the meeting.
8. Work with and approve the members of the Nominations Committee to identify candidates for elective offices and appointive committee memberships.
9. Review the By-laws and Officers' Manual, and when changes in these documents would improve the operation of the Society, work with the relevant officers to propose revisions for adoption by the Executive Committee.
10. The President can use discretionary funds, up to $500, to defray the President's Award winner cost to attend a national meeting (see President's Award).

Vice President/President-elect

1. Carry out duties assigned by the President and Executive Committee.
2. Contribute to site selection for annual symposia.
3. With Executive Committee approval, take over the duties of the President, should he/she be unable to serve.

Treasurer/Clerk

The Treasurer's term of office terminates not at the meeting in April, but prior to the Fall Executive Meeting. This allows business from the annual symposium to be settled before a new Treasurer takes office.

1. Present annual and semiannual reports on the NEAS Treasury (including Development Committee's monies) to the Executive Committee and to members of the Society at the general business meeting. The report should clearly indicate balances and interest payments/incomes on the accounts, receipts and expenditures.
2. File yearly income tax statements. These should be done by April 15th to ensurwe compliance with our Massachusetts Public Charities form.
3. Work with the Secretary/resident agent to ensure that the Massachusetts corporation's annual report is filed

each year by a Massachusetts resident with the Secretary of the Commonwealth of Massachusetts to-gether with the filing fee of $15. This maintains our status as a Massachusetts Corporation and allows us to be sales-tax exempt in Massachusetts.

4. Advise and insure that NEAS remains in compliance with its non-profit I.R.S. status.

5. Deposit all monies received from other officers or donors into NEAS bank accounts within two weeks of their receipt (see "Receipt of Monies" in Operating Manual).

6. Write all checks for the Society. Checks of greater than $1000.00 need a second approval (written or oral) by either the President or Secretary. Included in this are checks for the student travel awards and to the winners of the student presentation awards.

7. Work with the chair of the Development Committee to ensure that proper receipts for any donations to the society are mailed to donors in time for the tax forms.

8. Place all NEAS monies following payment of bills from annual symposium into an interest-bearing savings account; most of these funds should be put into a short-term, interest-bearing instrument (e.g. Certifi-cate of Deposit) in a timely fashion so that sufficient funds are accessible for payment of bills from No-vember until after payment of annual symposium expenses.

9. Provide sufficient cash to allow at the annual symposium to allow for making change at the development and society exhibits. If requested, you should also provide sufficient cash so that either the Treasurer or the co-conveners can tip the chefs and others after the banquet.

10. Prepare and distribute an IRS form 1099 misc to service providers that received cash from the society in time for their filing of income tax (generally by Feb 1 of the appropriate tax year).

11. Recommend improvements in investment and handling of monies to the Executive Committee (e.g. in con-sultation with local bank, etc.).

12. Transfer monies to a bank of the Treasurer-elect's choice within one month of the end of the term.

13. Advise President on any matters relating to the financial health of the organization. This generally includes reviewing all contracts for the annual symposium and informing the President of any irregularities in adherence to procedures for "Handling of NEAS monies."

14. Make sure that sufficient funds are taken off restricted instruments (e.g. Certificate of Deposit) in a timely fashion so that funds are accessible for payment of bills from November until after payment of annual symposium expenses.

Handling of NEAS monies

Officers and committee chairs who receive monies on behalf of the society (i.e. from members in payment for services, as donations to the society, etc.) shall transfer these monies to the Treasurer within two weeks of receipt unless other arrangements are made in advance. An adequate accounting of the monies shall be given to the Treasurer to ensure that they are deposited into the proper account and that the society can meet its IRS reporting requirements.

Co-conveners who will be receiving member's payments for the annual symposium should consult with the Treasurer in advance on how these funds are to be handled. Possible options include sending checks to the Treasurer to be deposited at regular intervals, holding the funds on site until the Treasurer can pick them up at the annual symposium, or depositing the funds in a separate temporary local account.

Credit card payments require special handling fees and handling requirements and need to worked out in advance. NEAS does have a Pay-Pal account and this option may be available. What option(s) will depend a lot on the infrastructure of the annual symposium, but the primary concern should be the safety of the funds and that NEAS can meet federal IRS reporting requirements.

Secretary

1. Handle archival materials of the society (committee reports, minutes of Executive Committee and general business meetings). At the conclusion of one's term, these documents should be passed to the next Secretary of NEAS.
2. Distribute copies of the Officer's Manual to all new officers. Incorporate changes to the Officer's Manual as approved by the Executive Committee.
3. Prepare and distribute an up-to-date list of contact information for all members of NEAS committees including the Development, Executive and ad hoc committees. This list should be distributed at least once per year and more often if the committee membership changes.
4. Write thank you notes to appropriate parties at the request of other officers (e.g. guest speakers, donors). Write a congratulatory letter to each award winner and a letter of thanks to retiring officers.
5. Work with the Treasurer and Massachusetts resident agent to ensure that the corporation's annual report is filed each year by a Massachusetts resident with the Secretary of the Commonwealth of Massachusetts together with the filing fee of $15. This maintains our status as a Massachusetts Corporation and allows us to be sales-excempt in Massachusetts.
6. Serve as Secretary at all Executive Committee meetings.
7. Prepare and distribute minutes of Executive Committee meetings to all committee members within one month of the meeting. These minutes should be reviewed and approved at the beginning of the subsequent Executive Committee meeting.
8. Prepare minutes of the general business meeting and distribute these to the Executive Committee members within one month of the annual symposium. These minutes should be offered for approval at the beginning of the following year's general business meeting.

Co-conveners

1. Consider possible symposium themes, mini-symposia, and prospective distinguished speakers. Investigate the costs associated with such plans.
2. Working with the staff at the designated symposium site, develop a list of projected costs for the symposium.
3. Report early arrangements, budget, logo, and general plans for the symposium to the Executive Committee at its autumn meeting.
4. Work closely with the NEAS Treasurer to decide who is going to receive funds, how funds will be transferred to the NEAS accounts, who is going to pay invoices and to ensure that the Treasurer has sufficient documentation to prepare the corporate annual reports and income tax statements. This should be done early in the process before any receipts and income are received.
5. Present the Executive Committee with a schedule for communicating with the membership, including distri-

bution of the first announcement, call for papers, registration information, deadlines for the abstract submission and registration, etc. Describe the planned means of communication.

6. With Executive Committee approval, invite the distinguished speaker.

7. Keep the President and other appropriate officers informed of progress in making arrangements.

8. At the earliest possible time, send a notice of the symposium date, location, theme, and invited speakers to the PSA Newletter.

9. Distribute the First Announcement (Membership Director supplies mailing labels if mail is used). The call for Nominations might also be included at this time. Send preliminary notices to Algae-L, and to the NEAS Webmaster for posting. The First Announcement should be in the hands of the membership by December 1, at the latest.

10. Distribute the Second Announcement, which will contain a formal call for registrations and payment of membership fees (i.e. payment for meeting program even if not attending), and the call for abstracts. State poster sizes and types of projection equipment available. Include Senior Phycologist Award solicitation. Include the Development Committee's call for student proposals for travel assistance, solicitation for items for the auction, travel information, map, etc. If the Call for Nominations was not distributed in an earlier communication, it should be included at this time. The Second Announcement and associated materials should be in the hands of the membership twelve weeks before the symposium. All announcements and forms should be provided to the NEAS Webmaster for posting on the NEAS website in a downloadable PDF file.

11. The deadline for abstracts and registration should be about six weeks before the symposium.

12. Explore whether a mechanism for arranging shared rooms is possible.

13. Send final reminder to Algae-L listserv at least two weeks in advance of the abstract deadline.

14. Solicit commercial exhibitors to bring displays of products to the symposium. List the eshibitors in the program. Arrange for space at the venue.

15. Provide the symposium logo to the Development Committee according to their schedule for producing items for sale at the annual symposium.

16. Assemble the scientific program. Determine the meeting schedule—opening, closing, time and location of breaks. Determine limits on schedules: wind-up time for banquet that gives staff reasonable time to clear up; deadline for key drop, etc.

17. Produce the symposium program with abstracts, biographical information and publication lists of the distinguished speaker or honorary chair, list of exhibitors, acknowledgements, etc. Provide an electronic version to the Membership Director for posting on the NEAS website.

18. Provide lists of needed equipment, registrants, meal/banquet attendees, number of vegetarians and others with special needs to the venue's conference manager.

19. Arrange for a room for the Executive Committee meeting. If the meeting is to be held during Saturday lunch, make arrangements that expedite prompt assembly and provision of lunches.

20. Prepare registration packages for attendees: name tags, receipts, meal and banquet tickets if applicable, general information about facilities, program with abstracts, election ballot. Identify a few places at which people arriving Friday night can eat and congregate.

21. Identify person to introduce the distinguished speaker.

22. Recruit session chairs.

23. Recruit registration desk helpers (students). Arrange for a table and chairs for registration workers.

24. Recruit projectionists. The incentive of a free banquet may be offered.

25. Arrange for a table and chair for use by the Development Committee.

26. Decide on the location of posters and commercial exhibits.

27. Recruit chairman of judging committee for student presentations; identify other potential members of this committee, provide chair with contact information for student members of the committee (Wilce and Trainor Award winners from the previous year). Provide a list of the students in each category. The correct listing of every student within each category is essential. The list should also be provided to the chair of the Development Committee, who will produce the award certificates.

28. After winners are determined, the Treasurer must be informed, and award checks written. The award checks and certificates should be provided to the President before the babquet for signing so that they may be presented at the banquet.

29. Prepare student name slips for door prize draws.

30. Ensure that prearranged honoraria and other payments are made to invited speakers; confirm that recipients of student travel awards and presentation award recipients have received their checks; and see that the venue has received all payments contracted for and that tips, if appropriate, have been distributed.

31. Work with the Treasurer to ensure that all residual funds are ransferred to the NEAS accounts and that all invoices from the annual symposium are paid in a timely manner (see Receipt of Monies).

32. Prepare a report to the Executive Committee for the archives, summarizing the number of registrants (separated by category), number of papers and posters, award winners, special lectures and sessions. Send the report to the Secretary, who will see that an appropriate version of this report is sent to the PSA Newsletter.

33. If the site at which the symposium was held has potential to be used again by NEAS, assemble a package in information, suggestions, and cautions that would expedite arrangements in the future.

Suggestions and observations:

Registration fees for 'Emeritus Members' should be about half the difference between students and professionals.

It is important that the social hour be long enough to accommodate the previous session running long, travel back to rooms to change clothes, and time to get to the site of the social. If necessary, plan a lengthened social hour to allow for all these variables plus one hour of socializing.

Membership Director

1. Maintain and update the NEAS mailing list in a form allowing for preparation of address labels.

2. Prepare printouts of mailing list to have at NEAS Executive Committee meetings and at the annual symposium for updating purposes.

3. Work with the Co-conveners to solicit interested individuals to add to the mailing list.

4. Maintain an e-mail address list of NEAS-related individuals categorized by status, e.g., undetgraduate students, graduate students, permanent professionals, transient professionals. Maintain a listing of governmental and commercial organizations with phycological interests.

5. Work with the webmaster to maintain the NEAS website. Solicit from the Executive Committee, symposium coneners, and chairs of standing and ad hoc committees information to be posted on the website.
6. Solicit from the Executive Committee, symposium conveners, and committees information to be disseminated to NEAS members or the larger phycological community by electronic mail.
7. Implement membership drives as directed by the Executive Committee.

Webmaster

The President will appoint a Webmaster to serve an indefinite term and that can be cancelled by the President or the Webmaster. The Webmaster will be a non-voting member of the Executive Committee.
1. Ensure that the NEAS website is up to date.
2. Work with the symposium co-conveners to promptly post any information regarding the upcoming annual symposium.
3. Work with the Treasurer to ensure that any subscription fees associated with hosting our website are promptly paid.
4. It is highly recommended that at least one other member of the NEAS officers has full permission and login information to gain access to the website in the event that the webmaster is unavailable. In the case of hosting a website at a university—this could include notifying the university IT personnel of who is the current NEAS President so that they can make changes if requested.

Chair of the Nominations Committee

1. The chair of the Nominations Committee is a voting member of the Executive Committee.
2. The chair appoints members of the Nominations Committee according to the Bylaws of the NEAS.
3. The chair provides to the co-conveners the Call for Nominations so that it can be sent out to the NEAS membership along with announcements of the NEAS spring symposium.
4. After nominations are received, the chair provides to the members of the Nominations Committee a list of candidates nominated for the offices to be filled and the number of nominations each received.
5. After consultation with the committee about the order in which the nominees will be solicited, the chair contacts the prospective candidates, asking them to serve.
6. The chair prepares the ballots along with biographical sketches solicited from the nominees and presented in a consistent format. This information will include previous service to NEAS.
7. The chair, aided by the Nominations Committee, oversees the election at the annual symposium.
8. Balloting and counting of ballots must be completed before the business meeting. The chair informs the President of the election outcome, and announces the winners as part of the Nominations Committee report at the business meeting.
9. The chair prepares a report on the final reslts of the election and the activity of the committee, and conveys the report to his/her successor.

Chair and members of the Lebaron Colt III Student Development Committee

The chair of the Development Committee is a voting member of the Executive Committee.

1. The goal of the Development Committee is to raise funds to support NEAS initiatives. Foremost among these initiatives is to provide financial assistance to student members, encouraging and enabling them to attend the annual symposium.

2. The chair of the Development Committee, along with the recipients of the Wilce and Trainor awards, shall serve as the student advocates at the meetings of the Executive Committee, ensuring that the needs of the students for attending the annual symposium are being met.

3. The chair or committee shall work with the co-conveners to decide on the mechanism for supporting students' attendance of the annual symposium.

4. The chair shall solicit requests for book awards from students and post-doctoral associates. These awards shall be approved by the committee as a whole and a list of successful winners forwarded to the Treasurer for purchase. Ideally this should occur at least one month before the annual symposium so that the books are available for distribution at the symposium banquet.

5. The Development Committee should work with the annual symposium conveners to coordinate the raising of funds to support student activities. This includes such activities as selling items with algal or NEAS logos (with approval of the Executive Committee), sponsoring auctions, and soliciting donations, such as through the Senior Phycologist Award. The chair of the Development Committee usually selects a person to serve as auctioneer during the banquet at the annual symposium and should arrange for staff to be present as part of the sales tables.

6. The chair of the Development Committee shall work closely with the Treasurer and the ad hoc student awards committee (see below) to ensure that support for society awards such as the Wilce Award for the best oral presentation and the Trainor Award for the best poster that are eligible for student to present their work at the annual meeting of the Phycological society is facilitated. In the case of disputes or requests for changes to the award, the chair of the Development Committee, along with the chair of the ad hoc committee, and the NEAS President shall arrive at a course of action so that the spirit of the award is achieved.

7. The society instituted the Senior Phycologist Award as a mechanism by which donations are solicited from senior professionals for the sole purpose of providing support to student recipients. The Development Committee shall select these student recipients and decide on the amount of the award. These awards should be approved by the President. The original recommendation was for each donor to provide $100. At the request of the donors, the past practice has been that the recipients are informed of the nature of the award, but not the name of the donors. Donors can be publically or privately acknowledged if they so desire.

8. The Development Committee should, where feasible, investigate other and new options for student support to advance education in phycology. This could include but is not limited to such new initiatives as student research awards, fellowships, internships, and other training opportunities.

9. The chair of the Development Committee should prepare an annual report detailing the above activities of the Development Committee for presentation at the spring Executive Committee meeting.

Responsibilities of the Publications Committee

The goal of the committee is to promote through publications broader understanding of the biological, ecological, and societal importance of the algae.

1. The Publications Committee shall recommend to the Executive Committee phycological studies and publications that might be sponsored by NEAS.
2. The Publications Committee will consider projects that NEAS might develop and sponsor, as well as proposals from outside the society. Such studies and publications may include floristic analyses, compilation of regional floras, posters, historical perspectives, economic and social issues dealing with the algae, and other studies that encourage an awareness and understanding of the importance of algae.
3. The chair of the Publications Committee should work closely with the Treasurer to ensure that any costs and profits associated with its activities are transferred in a timely manner to the appropriate NEAS accounts. The chair shall confer with the Treasurer and President about any costs associated with committee activities prio to incurrinfg those expenses.

Student awards and the ad hoc student awards committee

The Robert T. Wilce Award is given annually for the best graduate student oral presentation. Except in the case of extraordinary merit, no award will be made if fewer than five graduate student presentations are given. In cases of fewer than five participants, there will be no honorable mention in this category. The winner of the Wilce Award becomes a non-voting member of the Executive Committee for a one-year term. This student also becomes a member of the student award committee the ensuing year, and is therefore ineligible for an award in that year.

The Francis R. Trainor Award is given annually for the best graduate student poster presentation. Except in the case of extraordinary merit, no award will be made if fewer than five graduate student posters are given. In cases of fewer than five participants, there will be no honorable mention in this category. The winner of the Trainor Award becomes a non-voting member of the Executive Committee for a one-year term. This student also becomes a member of the student award committee the ensuing year, and is therefore ineligible to an award in that year.

The President's Award is for the best undergraduate student presentation (oral or poster). Except in cases of extraordinary merit, no award will be made if fewer than five undergraduate student presentations are given. In cases of fewer than five participants, there will be no honorable mention in this category. The President's Award winner can apply for additional funds to attend that year's PSA meeting (or another national or international meerting at the discretion of the President and the Executive Committee).

Award winners are selected by an ad hoc Student Awards Committee. Prior to the symposium, the conveners will select a chair of the student awards committee whose members will evaluate the student posters and oral presentations. If requested, the conveners will assist in recruiting members of the committee. The student awards committee should have at least three professional members evaluating contributions in each of the three award categories.

Professional awards and honors

The Frank Shipley Collins Award honors a member of NEAS who has contributed to the advancement of our society with exceptional effort and merit, as is embodied in the inscription of the award: "for the meritorious service to the Society and to Phycology."

The Executive Committee can receive nominations from any member of the Society. Awardees will be decided by vote of the Executive Committee, and the award will be made at the following annual symposium.

Incapacitation of an Officer

In the event of an illness or other unforeseen professional or personal problem that conflicts with the duties of his/her office, said officer shall notify the President as soon as possible. The President shall determine in discussions with that officer and the Executive Committee whether to seek additional or temporary help for this person or whether to replace the officer (see Bylaws). In the event of incapacitation of a co-convener, the conveners of the prior year will be the first potential replacement(s) or additional appointees. The period of time over which the conflict exists, its time of year, and the particular officer's duties will determine how serious the conflict is. It is clear that even a short (3 weeks) period of incapacitation of a convener between October-May might require the aid of the prior conveners in order to secure planning for the annual symposium.

Appendix D. *Conveners and members of the EC and DC up through 2010. Students not listed.*

Aaronson, Sheldon, Queen's College, Flushing, NY

Allen, Nina S., Dartmouth College, Hanover, NH

Anderson, Robert (Bob), Bigelow Laboratory for Ocean Sciences, West Boothbay Harbor, ME

Beale, Sam, Rockefeller University, New York, NY

Belsky, Melvin M., Brooklyn College, Brooklyn, NY

Bird, Carolyn J., National Research Council, Halifax, Canada

Blouin, Nicolas A. (Nic), University of Rhode Island, Kingston, RI

Boger, Edwin A. (Ed), Worcester State College, Worcester, MA

Bonneau, Ed, retired

Boyer, Gregory L. (Greg), SUNY, Syracuse, NY

Bradley, Peter, Worcester State College, Worcester, MA

Brawley, Susan H., Vanderbilt University, Nashville, TN; moved to the University of Maine, Orono, ME

Brinkhuis, Boudewijn (Bud) SUNY, Stony Brook, NY

Brody, Marcia, Hunter College, New York, NY

Cheney, Donald (Don), Northeastern University, Boston, MA

Chesnick, Joby, Lafayette College, Easton, PA

Chopin, Thierry, University of New Brunswick, Saint John, Canada

Coleman, Annette W., Brown University, Providence, RI

Colt, LeBaron (Barry), Southeastern Massachusetts University, North Dartmouth, MA

Cook, Philip (Phil), University of Vermont, Burlington, VT

Culver-Rymsza, Karen, University of Rhode Island, Kingston, RI

Domozych, David, Skidmore College, Saratoga Springs, NY

Fralick, Richard A. (Dick), Plymouth State College, Plymouth, NH

Gantt, Beth, Smithsonian Institution, Rockville, MD

Goff, Lynda J., University of California, Santa Cruz, CA

Golubic, Steve, Boston University, Boston, MA

Goos, Roger D., University of Rhode Island, Kingston, RI

Grobe, Carl, Westfield State College, Westfield, MA

Hanisak, M. Dennis, Harbor Branch Oceanographic Institute, Fort Pierce, FL

Hargraves, Paul E., University of Rhode Island, Narragansett, RI

Harlin, Marilyn M., University of Rhode Island, Kingston, RI

Heimke, John, Russell Sage College, Troy, NY

Heywood, Peter, Brown University, Providence, RI

Hillis-Colinvaux, Llewellya, Ohio State University, Columbus, OH

Holt, Jack, Susquehanna University, Selsingrove, PA

Hoops, Harold J., SUNY, Geneseo, NY

Johansen, H. William, (Bill), Clark University, Worcester, MA

Jones, Raymond F. (Ray), SUNY, Stony Brook, NY

Kepner, Raymond (Ray), Marist College, Poughkeepsie, NY

Klein, Anita, University of New Hampshire, Durham, NH

Laderman, Aimlee, Yale University, New Haven, CT

Lane, Christopher (Chris), Univeristy of Rhode Island, Kingston, RI

Lazaroff, Norman (Norm), SUNY, Binghamton, NY

Lee, Thomas F. (Tom), Saint Anselm College, Manchester, NH

Levandowsky, Michael (Mike) Pace University, New York, NY

Lewis, Louise, University of Connecticut, Storrs, CT

Liddle, Larry B., Long Island University, Southampton, NY

Lyman, Harvard, Brookhaven National Laboratory, Upton, NY

Mathieson, Arthur C. (Art), University of New Hampshire, Durham, NH

McManus, Hilary A., University of Michigan, Ann Arbor, MI

Muhlin, Jessica F., Maine Maritime Academy, Castine, ME

Neefus, Christopher (Chris), University of New Hampshire, Durham, NH

O'Kelly, Charles J., Bigelow Laboratory for Ocean Sciences, West Boothbay Harbor, ME

Page, Joanna Z., University of Connecticut, Storrs, CT

Parker, Henry S. (Hank), Southeastern Massachusetts University, North Dartmouth, MA

Peckol, Paulette, Smith College, Northampton, MA

Pintner, Irma, Pace University, New York, NY

Pueschel, Curt M., SUNY, Binghamton, NY

Ramus, Joseph (Joe), Yale University, New Haven, CT; moved to Duke University, Durham, NC

Robertson, Deborah, Clark University, Worcester, MA

Robinson, Beatrice B. (Bea), Le Moyne College, Syracuse, NY

Sager, Ruth, Hunter College, New York, NY

Saunders, Gary W., University of New Brunswick, Fredericton, Canada

Schmitter, Ruth E., Albion College, Albion, MI

Schneider, Craig W., Trinity College, Hartford, CT

Sears, James R. (Jim), Southeastern Massachusetts University, North Dartmouth, MA

Selsky, Melvyn I., Brooklyn College, Brooklyn, NY

Sheath, Robert (Bob), University of Guelph, Ontario

Siegelman, William (Bill), Brookhaven National Laboratory, Upton, NY

Siver, Peter A., (Pete), Western Connecticut State University, Danbury; moved to Connecticut College,
    New London, CT

Steneck, Robert S. (Bob), University of Maine, Walpole, ME

Swift, Dorothy G., University of Rhode Island, Kingston, RI

Sze, Philip (Phil), Georgetown University, Washington DC

Thursby, Glen B., U.S. Environmental Protection Agency, Narragansett, RI

Trainor, Francis R. (Frank), University of Connecticut, Storrs, CT

Vadas, Robert L., (Bob), University of Maine, Orono, ME

Van der Meer, John P., Atlantic Research Laboratory, Halifax, Canada

Van Patten, Margaret (Peg), University of Connecticut, Groton, CT

Villareal, Tracy, University of Texas, Port Aransas, TX
Vis, Morgan L., Ohio University, Athens, OH
Wehr, John, Fordham University, Armonk, NY
Wikfors, Gary, NOAA, NMFS, Milford, CT
Wilce, Robert T. (Bob), University of Massachusetts, Amherst, MA
Wysor, Brian, Roger Williams University, Bristol, RI
Yarish, Charles (Charlie), University of Connecticut, Stamford, CT

Bonus Section

Here is Charlie Yarish's acclaimed recipe for Blanc Mange pudding. Charlie says it may have originated with Jim Sears. Peg Van Patten likes it with Bailey's Irish Creme® and a slice of star fruit on top–a technique she observed in Dublin.

Blanc Mange Pudding

INGREDIENTS

1 quart milk
1 handful Irish Moss (1 cup)
3 Tablespoons sugar
1/2 teaspoon vanilla,
fruit or brandy, if desired

Collect *Chondrus crispus* (Irish moss) either as living plants from a rocky shore or as shorecast sprigs that have been bleached by sun and rain. Remove sand and small animals (they taste too much like the seashore), rinse briefly in fresh water, and either use immediately or dry for later use. Dry moss can be stored indefinitely in a dry, airtight container.

Simmer milk in a double boiler (or use a single pan, but be careful not to burn the milk). Add seaweed to hot milk and simmer for about 20 minutes, until viscous. Carrageens extracted from the moss increase the viscosity of the pudding, causing it to gel. Remove the seaweed from the pudding and place it on your compost pile, or leave it in to add texture. Save the viscous milk. Sweeten and flavor with sugar and vanilla. Add fruit or brandy for variety. Set aside to gel.

Photo Gallery

Some randomly-snapped informal photos from various meetings over the years.

Author Peter M. Bradley and wife Rosslyn enjoy a NEAS meeting.

Craig Schneider, Bob Wilce and author Bill Johansen.

Katy Hind (U. New Brunswick), Bill Johansen, and Jim Sears.

George Papenfuss, distinguished speaker at the 1977 meeting.

Ed Moul

Long-time NEAS auctioneer, Glen Thursby, in 2009

Susan Brawley, editor of Journal of Phycology for many years.